SPACE TRAFFIC MANAGER'S HANDBOOK

THE ASPIRING

SPACE TRAFFIC MANAGER'S HANDBOOK

FROM SPACE OBJECTS TO SPACE DEBRIS

Cristian Bulumac

Bruxelles, 2024

space@bulumac.ro

www.space.bulumac.ro

Legal deposit: D/2024/Cristian Bulumac, editor

The Aspiring Space Traffic Manager's Handbook

Cristian Bulumac - 1st ed.

ISBN 978-2-9603408-0-8 (paperback)

ISBN 978-2-9603408-1-5 (ebook)

Do not look at stars as bright spots only. Try to take in the vastness of the universe.

—MARIA MITCHELL 1818-1889, ASTRONOMER

Contents

CHAPTER ONE

Introduction

If you look up at the night sky and are fortunate enough to be in an area with less light pollution, the countless stars twinkling in the vast expanse of space will captivate you. However, the roads humans and their machines use to travel amongst the planets and stars are invisible within the celestial beauty. Many of us dream that one day, humans will be able to explore the far depths of outer space. We hope that one day we will become a "multiplanetary species", and with each passing day, the children's dream to become astronauts is becoming more viable. Over the past century, humanity has made tremendous strides in space exploration. The best is yet to come, but it is challenging.

The New Frontier is not the new Wild West; space debris could be one of the most significant setbacks when leaving Earth's atmosphere. In 2022, more satellites than ever were deployed to orbit the Earth, which is impressive. However, many of these objects will remain in Earth's orbit well beyond their expiration date, and so they pose a significant collision risk to future missions.

We tend to use the rarity of being hit by lightning as an example of an extraordinary occurrence. However, in 1997, a person was hit by space debris. What are the chances of that? This incident is the first one we know of, but an average of one known piece of debris has fallen back to Earth each day for the past 50 years, and we have just begun to send stuff up there.

The first policy proposals around space management date back to the early 1930s, and the world is ever-changing. Today, we need an updated policy and approach for optimal cooperation between industry and governments. This will result in a sustainable path forward for future missions.

The United Nations Office for Outer Space Affairs (UNOOSA) reported 11,330 satellites orbiting the Earth as of June 2023. This represents a staggering 37.94% increase since January 2022, and, in terms of launching, 36.68% more than the previous record of 1,810 items launched in 2021. An exponential trend is occurring as 17 planned networks comprised of thousands of satellites each, the famous mega-constellations, are announced, totalling 536,301 planned satellites to share the limited space around Earth.

In the vastness of outer space, even the tiniest speck of paint can become a destructive force hurtling at a mind-boggling velocity of 25,000 km/hour. Incidents have already happened and left their mark on spaceships and satellites crucial for weather forecasting, traffic control, geolocation, navigation, broadcasting, communication, and the internet.

As societies increasingly rely on satellites for daily operations, the Earth's orbital space, often referred to as a common-pool resource, faces unprecedented challenges.

The treaties governing outer space, notably the 1967 Outer Space Treaty, declare that outer space "shall be free for exploration and use by all States." This means that activities that prevent others from enjoying the benefits of Earth's orbits are a problem. Furthermore, the competitive nature of these orbits implies that once a spacecraft or debris occupies a specific orbital slot, another cannot simultaneously take its place. If two objects try to occupy the same space, the people in the industry say that a "rapid unscheduled disassembly" follows. People like you and me would say that those two objects will shatter to pieces, thus creating more objects and increasing the risk of more collisions.

This is especially true in heavily congested orbits, turning them into valuable but challenging common-pool resources.

But what happens when the space belongs to everybody and nobody in particular? The "tragedy of the commons" is a concept in economics and environmental science. It shows how a shared resource can be depleted or degraded when individuals exploit it without considering the common good. This leads to overuse, depletion, or pollution of the resource, causing negative consequences for everyone involved. Earth's orbital space is especially vulnerable to this. In the context of space, overexploitation without investing in conservation leads to congestion and an abundance of debris, posing a significant threat to sustainable space activities.

Technological advancements and increased scarcity have amplified the challenges faced by Earth's orbital space. Lower launch costs and the development of smaller satellites have democratized access to space. More countries and private entities are now launching satellites. The Outer Space Treaty envisions space exploration as a province of all humankind, calling for a shared responsibility over this celestial domain.

However, despite the urgent need for sustainable practices, few measures have been implemented, and the tragedy of space debris continues to escalate. Let's unravel this one step at a time.

Buckle up; we have liftoff in 3... 2... 1....

CHAPTER TWO

Milestones in human space exploration

In human history, exploring space is one of our most audacious and awe-inspiring endeavours. From dreaming of escaping Earth's gravity to humans landing on the Moon and planning bases on the Moon and Mars, it shows our creativity, determination, and curiosity.

The seeds of space exploration were sown at the turn of the 20th century. Visionaries like the Russian schoolteacher Konstantin Tsiolkovsky and the American physicist and inventor Robert H. Goddard have taken the ancient firework propulsion and conceptualized the principles of rocketry, laying the groundwork for humanity's first tentative steps beyond the Earth's atmosphere.

Born from the crucible of war, the V-2 rocket, developed by Nazi Germany during World War II, unwittingly became the launchpad for the United States' journey into the cosmos and an inspiration for other countries' rocketry technology.

After the war, as part of Operation Paperclip, a group of German scientists, including Wernher von Braun, were brought to the United States. The Redstone and Jupiter rockets, descendants of the V-2, became foundational elements of early American space endeavours, setting the stage for subsequent milestones such as the Mercury and Gemini programs and, ultimately, the Apollo missions that landed humans on the Moon.

The Cold War rivalry between the United States and the Soviet Union, which started after the Second World War, fueled the race to go further in space.

On 4 October 1957, the Soviet Union launched Sputnik 1, the world's first artificial satellite. This momentous event marked the beginning of the Space Age and ignited the intense competition known as the Space Race.

Yuri Gagarin's historic orbit around Earth in 1961 made him the first human in space, and the United States responded with the Apollo program. The culmination of this endeavour was the iconic Apollo 11 mission, where astronauts Neil Armstrong and Buzz Aldrin became the first humans to set foot on the lunar surface on 20 July 1969.

The 1980s saw the advent of the Space Shuttle program, promising reusable access to space. Shuttles like the Challenger and Discovery became household names as they ferried astronauts and payloads into orbit. The Hubble Space Telescope, deployed in 1990, provided unprecedented views of distant galaxies, expanding our understanding of the cosmos.

As the Cold War thawed, international cooperation blossomed in space. The Soviet Union's Mir space station, launched in 1986, hosted cosmonauts and astronauts from various nations.

Mir was the first continuously inhabited long-term research station in orbit. It held the record for the longest continuous human presence in space at 3,644 days until it was surpassed by the International Space Station on 23 October 2010.

Mir's legacy endured in the International Space Station, a multinational endeavour that began construction in 1998 and continues to serve as a microgravity laboratory and symbol of global collaboration.

While humans explored space, robotic emissaries ventured where no human could yet go. The first interplanetary missions, like the Soviet Union's

Luna 1 in 1959 and the United States' Mariner 2 in 1962, laid the foundation for planetary exploration.

In 1977, the Voyager program saw the launch of two robotic interstellar probes, Voyager 1 and Voyager 2. As of 2024, the Voyagers are still in operation beyond the outer boundary of the heliosphere in interstellar space.

Other landmark missions include the Mars rovers Spirit and Opportunity, which touched down on the Red Planet in 2004, and the Juno spacecraft, which entered orbit around Jupiter in 2016.

The 21st century heralded a new era of space exploration. In 2006, the New Horizons spacecraft performed a historic flyby of Pluto, capturing detailed images of this distant dwarf planet.

Meanwhile, private companies like SpaceX, founded by Elon Musk in 2002, began making strides in commercial spaceflight, launching satellites and cargo to the ISS and pioneering reusable rocket technology.

Notably, the countries in Asia emerged as significant players in space exploration. India's Indian Space Research Organisation (ISRO) achieved important milestones with its Mars Orbiter Mission (Mangalyaan) in 2013, making India the first Asian nation to reach Martian orbit and showcasing its technological prowess. Japan's JAXA (Japan

Aerospace Exploration Agency) also contributed to our understanding of the cosmos with missions like the Hayabusa series, which successfully returned samples from asteroids.

One of the noteworthy additions to the international space scene was the emergence of China as a significant player. China's space program, marked by achievements like the Shenzhou missions that carried humans into space, reached a significant milestone with the construction of the Chinese space station, Tiangong ("Heavenly Palace"). Launched in modular components, Tiangong is still growing through the planned addition of new modules.

Humanity stands today at the threshold of exciting possibilities, with plans for crewed missions to Mars, permanent bases on the Moon, continued robotic exploration of the outer planets, and the expanding potential for commercial space activities and tourism.

Before shooting for the stars, humankind created space objects that went round and round the Earth, sometimes moving towards the edges of our solar system. It is important to understand what they were and how they evolved to share the space around our planet.

CHAPTER THREE

Space Objects Chronicles

A satellite can simply be defined as an object that moves around a larger object on a pathway called orbit.

The idea of artificial satellites traces back to the early 20th century when science pioneers envisioned the possibility of sending objects into orbit around Earth. Their visionary writings and calculations drew upon a plethora of forbearing scientists, like Aristarchus of Samos, Erathostenes, Tycho Brahe, Johannes Kepler, and Sir Isaac Newton. Their work laid the foundation for the scientific understanding of space travel and the concept of satellites.

One of the key figures in the development of satellite technology was Arthur C. Clarke, a British award-winning science fiction writer and scientist.

In a 1945 paper titled "Extra-Terrestrial Relays," Clarke proposed the concept of using geostationary satellites for global communication.

He recognized that by placing satellites in orbits at an altitude of approximately 36,000 kilometres (22,000 miles) above Earth's equator, they would appear stationary from the ground, allowing for continuous communication coverage over large areas.

Building upon these visionary ideas, engineers and scientists began to explore the practicalities of satellite development. During the Cold War space race between the United States and the Soviet Union, the stage was set for a groundbreaking moment in human history.

The Soviet Union stunned the world by successfully launching Sputnik 1, the first artificial satellite. This basketball-sized sphere emitted a distinct beeping sound as it orbited Earth, capturing the imagination and awe of people worldwide. Sputnik 1 remained in orbit for about three months, transmitting valuable data about Earth's upper atmosphere density and the effects of atmospheric drag on satellites.

The launch of Sputnik 1 not only marked a significant technological achievement but also triggered a profound shift in geopolitics. It ignited the space race between the United States and the Soviet

Union, as both nations rushed to demonstrate their technological prowess and assert their dominance in space exploration.

Following Sputnik 1, numerous satellites were launched by both superpowers, advancing scientific knowledge and enabling groundbreaking discoveries. Notable missions included the Soviet Union's Luna program, which sent probes to the Moon, and the United States' Explorer program, which investigated Earth's magnetosphere and cosmic rays.

The dawn of the Space Age brought remarkable advancements in satellite technology.

Miniaturization of electronic components, improvements in propulsion systems, and the development of solar panels for power generation enabled satellites to become more sophisticated and capable.

The launch of Telstar 1 in 1962, the world's first commercial communication satellite, ushered in a new era of global connectivity, enabling live television broadcasts and international telephone calls.

The subsequent decades witnessed an exponential growth in satellite applications and capabilities. Satellites became crucial tools for Earth observation, weather forecasting, navigation, scientific research, and telecommunications. The deployment of satellite constellations, such as the Global Positioning System

(GPS) and the Landsat program, revolutionized industries and transformed our understanding of the world. Since then, many more countries joined the race and launched satellites and probes.

Most of these are sharing a space called Low Earth Orbit. If we compare our planet and its surroundings to an orange, Earth is like the edible core, and the white part of the peel is the atmosphere. The area starting with the very thin, orange-coloured part of the peel is the Low Earth Orbit layer just beyond the atmosphere. This is where humankind has launched most of its space objects.

Today, satellites have become an integral part of our modern society, facilitating global communication, providing critical data for environmental monitoring, enhancing navigation systems, and opening new frontiers in scientific exploration. They continue to evolve, with technological advancements enabling smaller, more powerful, and more agile satellites, paving the way for future space research and exploration breakthroughs.

But satellite is such a versatile word, sometimes used in songs to express attraction towards someone and in science to describe various objects with different characteristics and purposes.

So, instead of going around the topic like a satellite, you can find below some of the most important satellites and categories, ranging from space stations (yes, they are satellites, too) to their cousins at the edge of the solar system, the deep space probes.

Salyut Space Stations

Salyut 1 made its mark as the world's first space station. The Soviet Union launched the 18,900 kg spacecraft into Low Earth Orbit on 19 April 1971, making history. Salyut 1 was not without its challenges, but it was a great success for the Soviets in many ways. It made 2,929 orbits of Earth over 175 days, and in 23 of these days, it was staffed. During its lifespan, it covered over 118 million kilometres in distance. Every 88.5 minutes, Salyut 1 would make an entire revolution of the Earth at an altitude of about 210km.

The first ever crew on the first space station arrived with the two-person Soyuz 11. They successfully docked and performed experiments in Salyut 1 for 23 days. Sadly, both of them were killed by asphyxia caused by the failure of a valve just before reentry. This makes them the only people to have died above the Kármán line, the line known but not universally accepted as the boundary of space.

Salyut 1 re-entered the Earth's atmosphere on 11 October 1971 and burnt up.

Salyut 1 was a modified version of the Soviet military's then-under-development Almaz space station program. The Soviets redirected the priority of their crewed space program after the landing of Apollo 11 on the Moon in July 1969 to orbiting space stations, with the possibility of a lunar landing in the later 1970s if the N-1 rocket proved safe and functional. Although the first six civilian Soviet space stations were referred to as Salyut, the Russian name is DOS, a Russian acronym that translates to "long-duration orbital station".

The craft's dimensions were substantial, measuring 20 meters in length, 4 meters in maximum diameter, and containing 99 cubic metres (3,500 cubic ft) of internal space. It featured multiple compartments, including three pressurized ones, two of which were accessible to the crew.

A notable fact about Salyut missions is that in 1984, Svetlana Savitskaya became the first woman to exit the station and perform activities in the open space, also known as a spacewalk. The first man to ever spacewalk was Alexei Leonov, who did so in 1965 during the Voshkod-2 mission. The experience was unusual because once he exited the pressurized atmosphere of the spaceship, the space suit overinflated, making him unable to fit through the door and return until he let some air out.

The Voshkod-2 mission, like all missions in the dawns of space exploration had a lot of troubles, but let's get back to Salyut.

The interior design of Salyut 1 included a range of hues, including light yellow, apple green, and dark and light grey, to help the cosmonauts' orientation in weightlessness. Eight large seats (seven at workstations), many control panels, and twenty portholes (some obscured by instrumentation) were visible during televised views. Beyond this, there were a number of extra features, including a life support system, power supply, control, and communication devices, all housed within a third pressurized compartment. The engine installations and related control equipment were housed in an unpressurized compartment.

A telecope named Orion 1 Space Observatory, designed by the Armenian astronomer Grigor Gurzadyan, was a part of Salyut 1. The telescope took ultraviolet spectrographs of stars, and Viktor Patsayev, a crew member, operated the telescope, making history as the first person to command a telescope outside of Earth's atmosphere.

Following the Salyut program, several additional stations used the lessons learned and the technology. Zvezda (DOS-8), the program's last module, is still in orbit and serves as the foundation for the Russian sector of the International Space Station.

Skylab Space Station

The United States' reply to Salyut was Skylab, the first space station to be crewed by the United States. The station was occupied for around 24 weeks, spanning from May 1973 until February 1974, by three different crews of three astronauts. After Skylab's orbit began to deteriorate, on 11 July 1979, it re-entered and broke apart in the atmosphere, sending debris over Western Australia and throughout the Indian Ocean.

With the 14,000-kg (31,000-pound) Apollo command and service module (CSM) connected, Skylab weighed 90,610 kg (199,750 pounds). It also had several hundred physical and biological research experiments, a workshop, and a Sun observatory.

The orbital workshop was constructed from the S-IVB rocket's third stage, and a more compact Saturn IB rocket was used to launch three astronaut crews on Apollo Command and Service Module flights. The primary liveable area of Skylab was the orbital workshop.

Other features of Skylab were the Apollo Telescope Mount, a multispectral solar observatory, an airlock module with extravehicular activity (EVA) doors, and a multiple docking adaptor with two docking ports.

The Apollo Command and Service Module was docked to it and used fuel cells and solar arrays to generate electricity. There was a heat radiator, propellant tanks for the manoeuvring jets, and a giant waste tank in the back of the station.

Astronauts used Skylab for many experiments during its existence, but the space station was also at the core of an interesting predicament: when pieces of Skylab fell to Earth in 1979, debris landed in a place in Australia called Esperance. The Shire of Esperance sent NASA a $400 littering fine, but NASA chose to ignore it, and the fine was paid 30 years later on their behalf by a radio host from California.

Mir

Mir, which translates from Russian as "peace" or "world," was a Low Earth Orbit space station operated by the Soviet Union and after that by Russia from 1986 until 2001.

Mir's construction between 1986 and 1996 marked a significant leap in space technology, featuring multiple interconnected modules that served various purposes. Over its operational life, Mir hosted cosmonauts, astronauts, and international visitors, becoming a symbol of cooperation during the post-Cold War era.

It weighed more than any previous spaceship. It was the biggest artificial satellite in orbit at the time, although the International Space Station (ISS) surpassed it later. The station acted as a microgravity research laboratory, where crews performed research in cosmology, physics, and human biology to create technologies necessary for long-term space exploration.

Mir was the first continuously inhabited long term research station in orbit, and it maintained the title for the longest continuous occupation by people in space at 3,644 days until it was surpassed by the International Space Station on 23 October 2010. Mir was inhabited for twelve and a half years of its fifteen-year existence, with the ability to host a three-crew mission or even bigger teams for brief visits.

Mir had multiple modules that were added over time. Modules included living quarters, laboratories, and docking ports. When finished, the station had seven pressurized modules and various unpressurized components. Several solar arrays, connected directly to the modules, produced power. The station was kept in an orbit between 296 km and 421 km above the Earth's surface, travelling at an average speed of 27,700 km/h and completing 15.7 orbits daily. It weighed about 130,000 kg and was approximately 20 meters in length.

MIR was also the scene of a space accident. On 25th June 1997, an error of the control and approach system caused a collision between the Progress M-34 transport ship and the docked Spektr module of the Mir station. The damaged module was sealed off and remained attached until the station was brought down.

Mir ended its mission on 23 March 2001, when it was brought out of its orbit, entered the atmosphere and was destroyed. Given the evolution of technology, millions of people witnessed its destruction and documented it in videos and photographs. The South Pacific Ocean is the final resting place of the debris that survived reentry.

The collaborative efforts on Mir helped build trust and paved the way for the extensive international partnerships that continue to collaborate with the International Space Station, where agencies from the United States, Russia, Europe, Japan, and Canada continue to work together on scientific research and space exploration.

The International Space Station

The International Space Station (ISS) is an orbiting laboratory that has housed over 250 people since 1998. It is the largest single structure ever placed in space by humankind. The station's major construction took place between 1998 and 2011; however, it is constantly evolving to incorporate new endeavours and experiments. It has been occupied continuously since 2 November 2000.

Excluding visiting vehicles, the International Space Station spans 109 meters in length and weighs 419,725 kilograms. The station's solar arrays cover 2,500 square meters (27,000 square feet).

Crew members have 13,696 cubic feet of liveable capacity. The space station contains seven sleeping quarters, two restrooms, a gym, and a cupola, which provides a 360-degree view of Earth. Astronauts have often compared the living quarters of the space station to the cabin of a Boeing 747 aeroplane.

The International Space Station orbits the Earth at an altitude of around 400km, periodically every 90 minutes at an average speed of approximately 28,000 km/h (17,500 mph). The station traverses the distance between Earth and the Moon in a single day.

As of May 2022, 258 people from 20 nations had travelled to the International Space Station. The top two nations in terms of participation are the United States (158 participants) and Russia (54 participants). Space agencies are assigned places and research time on the space station based on the quantity of financial resources they provide.

The International Space Station is funded by 15 different countries. The space station's principal collaborators and funders are NASA (United States), Roscosmos (Russia), and the European Space Agency (ESA); additional partners include the Canadian Space Agency and Japan Aerospace Exploration Agency (JAXA). Commercial astronauts are beginning to operate on the orbiting complex through a commercial corporation named Axiom Space; also, astronauts from other countries occasionally participate in missions to the space station.

The space station is scheduled to be operational until at least 2024. NASA has granted an extension until 2030, but Russia has stated that it would leave after 2024 to focus on creating a space station of its own around 2028. The station's operation following Russia's departure has yet to be resolved, and the objectives for the International Space Station after 2030 are also unclear. It might be deorbited or recycled for use in future space stations.

The International Space Station is noticeable from Earth at night, appearing as a blazing floating point of light that rivals the brilliance of the magnificent planet Venus.

Currently, SpaceX's Crew Dragon spacecraft or, in the instances involving Russian participants, the Soyuz capsules transport people to the space station. After NASA's space shuttle program ended in 2011, the Soyuz became the principal mode of travel for all those involved. Crew Dragon started transporting humans with the Demo-2 mission, which took off on 30th May 2020.

Space debris forced the International Space Station to perform evasive manoeuvres dozens of times. A December 2022 NASA report states that since 1999, the International Space Station has course-corrected itself 32 times to avoid satellites and trackable space debris.

Tiangong Space Station

Tiangong — which translates to "Heavenly Palace" — was developed in Low Earth Orbit by the Chinese Manned Space Agency (CMSA), with each of the station's three modules launched between 2021 and 2022. The CMSA deployed Tianhe, the first station module, on 28 April 2021, Wentian, the second module, on 24 July 2022, and Mengtian, the third module, on 31 October 2022.

Shenzhou 12, China's first crew to visit the space station, arrived on 16 June 2021, when only the core module Tianhe was in orbit. The taikonauts – the name used by Westerners for Chinese astronauts — stayed in Tianhe for 90 days, about three times the duration of any prior Chinese crewed spacecraft.

The CMSA intends to maintain Tiangong occupied by at least three crew members for at least a decade. During that time, the space station will be home to several experiments from China and other countries.

Tiangong circles Earth at an altitude of 340 to 450 km (217 to 280 miles), almost the same as the International Space Station.

Although Tiangong is substantially smaller in size than the International Space Station, having only three modules as opposed to the ISS's 16 modules, it is also far lighter, weighing about 20% of ISS's mass. The Tianhe module, which is 16.6 meters long, is fitted with a docking hub, enabling it to accommodate Shenzhou crew and Tianzhou cargo ships. A robotic arm assisted astronauts during spacewalks and helped place the Mengtian and Wentian modules.

On the other hand, Tianhe is larger than the nation's Tiangong 1 and 2 experimental space labs flown during the previous decade, weighing in at 24 tons. With the new Tiangong and the connected

Shenzhou and Tianzhou vessels, China's taikonauts have a lot of usable room.

Tiangong will be accompanied by a massive space telescope, similar to the famous Hubble Space Telescope, which will be able to dock with the space station for repairs, maintenance, and potential upgrades. The Xuntian telescope, meant to "survey the heavens," will feature a 2-meter-wide mirror somewhat smaller than Hubble's; however, it boasts a field of vision 300 times greater. Using its massive, 2.5-billion-pixel camera, Xuntian hopes to study 40% of the sky in ten years. Some people have expressed concerns that the station arm and the telescope might be used for other purposes, but that is neither surprising nor unprecedented.

If all goes as planned, the space station might be upgraded to include six modules. The addition of two more modules to the orbiting outpost might be enabled by the second Tianhe core module.

CSMA has focused on creating new, more expansive rockets to enable the construction of a space station. The Long March 5B was particularly built to launch the massive space stations through Low Earth Orbits. Still, the rocket was the cause of one of the greatest uncontrolled reentries in recent decades, fortunately without negative consequences. Mengtian and Wentian launches exhibited similarities to the uncontrolled Long March 5B crashes, garnering more condemnation from some countries. In recent years, China completed its new maritime spaceport at Wenchang, primarily for carrying these larger diameter rockets that must be supplied by sea.

Communication Satellites

Communication satellites are space objects found primarily on Low Earth Orbits and Geostationary Orbits, facilitating the transmission and reception of signals for various communication services. They act as "middlemen" between ground based transmitters and receivers, enabling the seamless exchange of information across continents and oceans.

At the heart of satellite communication is the principle of signal transmission via radio waves. Ground-based transmitters encode information, such as voice, data, or video, into radio frequency signals. These signals are then beamed up to communication satellites in space.

Communication satellites are equipped with transponders, devices that receive incoming signals, amplify them, and then retransmit them to Earth. Transponders operate within specific frequency bands, allocating different frequencies for uplink (from Earth to the satellite) and downlink (from the satellite to Earth) transmissions.

When a signal reaches a communication satellite, the onboard transponder receives processes and amplifies it to ensure a strong and clear transmission. The amplified signal is then retransmitted back to Earth, where it can be received by ground-based antennas.

The coverage area of a satellite is known as its footprint. The footprint of a satellite can cover an entire continent or a significant portion of the Earth's surface, depending on the design and purpose of the satellite. Multiple communication satellites strategically positioned around the globe create a network that ensures global coverage.

The Geostationary Orbit, in particular, is a highly sought-after location for communication satellites due to its unique advantages. By placing satellites in this orbit, they remain fixed in the sky relative to an observer on the ground. This characteristic allows users to establish a connection with the satellite using fixed antennas without the need for constant repositioning.

There are many types of what we can refer to broadly as communication satellites, and most ones fit in one of the following major fields:

1. Telecommunication satellites, connect people worldwide, facilitate long-distance phone calls, and support voice-over-IP (VoIP) services and video conferencing, enabling real-time communication and collaboration on a global scale.

2. Broadcasting satellites facilitate the transmission of television and radio signals across continents. Direct-to-home (DTH) television services rely on communication satellites to deliver various channels to households worldwide. Satellite radio services provide uninterrupted music and news broadcasts, offering listeners diverse programming options.

3. Internet Connectivity satellites bridge the digital divide by providing connectivity, in particular to remote and underserved regions. Satellite internet services deliver internet access to rural areas, ships, aeroplanes, and other locations where terrestrial infrastructure is limited or absent. They contribute to the global effort of expanding digital inclusion and fostering equal access to information and opportunities. Beyond the existing GEO high latency satellites, several mega-constellations of LEO satellites are currently planned for this purpose, with Starlink and One Web being the first active ones.

4. Disaster Response and Emergency Communications satellites play a vital role in

disaster management and emergency response. Ground-based communication infrastructure may be damaged or unavailable during natural disasters or humanitarian crises. Satellites provide reliable and rapid communication links for emergency services, enabling coordination, information dissemination, and assistance to affected areas.

Remote Sensing Satellites

Remote sensing satellites enable us to gather information about Earth from a once unimaginable perspective. By utilizing a variety of sensors and imaging systems, these satellites capture data across the electromagnetic spectrum, providing us with an unprecedented view of our planet.

One of the key technologies employed by remote-sensing satellites is optical imaging. These satellites utilize cameras equipped with sophisticated lenses and detectors to capture high-resolution images of the Earth's surface. They can differentiate between various land covers, monitor changes in vegetation patterns, and track the expansion of urban areas. Optical sensors can capture images in different wavelengths, including visible, near-infrared, and thermal infrared, allowing us to gather crucial information about Earth's processes and dynamics.

Furthermore, remote sensing satellites employ radar imaging systems. Synthetic Aperture Radar (SAR) is a widely used technology that uses radar signals to map the Earth's surface. By measuring the time it takes for the radar pulses to return to the satellite, SAR can create detailed images regardless of weather conditions or the time of day. This capability makes SAR particularly useful for monitoring forests, mapping topography, and detecting changes in the Earth's surface, such as subsidence or the movement of glaciers.

Another vital technology utilized by remote sensing satellites is multispectral imaging. By capturing images in multiple spectral bands, these satellites provide a wealth of information about Earth's features and properties. Multispectral sensors can identify specific wavelengths corresponding to various materials, enabling the discrimination of different vegetation types, identification of minerals, and detection of pollutants in water bodies. These sensors contribute to our understanding of environmental changes, land use patterns, and the health of ecosystems.

Remote sensing satellites also play a crucial role in monitoring Earth's atmosphere and weather patterns. They carry instruments such as radiometers and spectrometers that measure the atmosphere's composition, temperature, and dynamics. These measurements are vital for weather forecasting, climate studies, and monitoring atmospheric pollutants and greenhouse gases. Satellites equipped with atmospheric sensors provide real-time data on weather systems, enabling meteorologists to issue accurate forecasts and warnings, ultimately saving lives and mitigating the impacts of severe weather events.

Furthermore, remote sensing satellites contribute to studying Earth's oceans and coastal regions. They carry sensors that can detect and measure properties such as sea surface temperature, ocean colour, chlorophyll concentration and other marine constituents. These measurements aid in the monitoring of oceanic currents, the study of marine ecosystems, and the detection of harmful algal blooms. Remote sensing satellites provide a comprehensive view of our vast oceans, helping scientists understand their dynamics, track changes, and promote the sustainable management of marine resources.

The data gathered by remote sensing satellites is valuable for scientific research and has practical applications in various sectors. It aids land-use planning, agricultural management, disaster response, and resource exploration.

Satellite imagery enables us to assess the impact of natural disasters, monitor deforestation, track the spread of wildfires, and manage water resources more efficiently. The insights derived from remote sensing data enhance decision-making processes, promote sustainable development, and contribute to the preservation of our planet.

In recent years, the advancement of remote sensing satellites has been accompanied by the emergence of constellations of small satellites, enabling more frequent and targeted data acquisition. These constellations offer the potential for near real-time monitoring and increased spatial resolution, revolutionizing our ability to study Earth's dynamic processes and address pressing environmental challenges.

The so-called spy satellites are one particular case of remote sensing satellites, as they monitor, record and transmit data for intelligence or military use.

Navigation Satellites

Navigation satellites play a crucial role in providing accurate positioning, timing, and navigation services worldwide. These satellite-based systems utilize a constellation of satellites, ground based receivers, and sophisticated algorithms to determine precise locations on Earth's surface.

To determine their position on Earth, satellite-navigation receivers on the ground collect signals from multiple satellites simultaneously. The receiver can calculate its distance from each satellite by measuring the time it takes for the signals to reach the receiver and the precise timing information that the satellites provide. Using trilateration, a geometric method, the receiver can determine its precise location on Earth's surface.

Satellite-navigation offers exceptional accuracy, with modern receivers capable of determining positions with an accuracy of a few meters or even centimetres. This level of precision has revolutionized navigation in various sectors, including aviation, maritime transportation, surveying, and personal navigation, as receivers can now fit on a chip in mobile phones.

One of the most prominent satellite-based navigation systems is the Global Positioning System (GPS), initially developed by the United States Department of Defense for military purposes. GPS comprises a network of satellites orbiting approximately 20,000 kilometres (12,000 miles) above Earth. These satellites continuously transmit signals that contain precise timing information and their orbital parameters.

Beyond GPS, other satellite-based navigation systems have emerged to provide alternative or complementary services. Notable examples include the Russian GLONASS (Global Navigation Satellite System), the European Galileo, and the Chinese BeiDou Navigation Satellite System. These systems operate on similar principles to GPS, utilizing constellations of satellites to provide accurate positioning and navigation information. Many devices can receive signals from more than one satellite constellation, thus improving precision.

The applications of satellite-based navigation systems are vast and varied, with some notable examples below:

1. Navigation satellites have transformed air travel by enabling precise navigation, approach, and landing procedures. Pilots rely on satellite navigation systems to follow designated routes, avoid obstacles, and conduct accurate instrument approaches, enhancing safety and efficiency in the skies.

2. Navigation satellites are essential for maritime transportation, providing ship captains and navigators with accurate position and course information. Satellite-based navigation systems enable efficient route planning, collision avoidance, and precise docking manoeuvres, contributing to safer and more reliable maritime operations.

3. Satellite navigation systems have become integral to personal navigation devices like smartphones, car navigation systems, and wearable devices. These devices utilize satellite signals to provide turn-by-turn directions, real-time traffic updates, and location-based services, making it easier for individuals to navigate unfamiliar areas and reach their destinations.

4. Satellite-based navigation systems revolutionize the field of surveying and mapping by providing precise positioning information. Surveyors can accurately measure distances, map terrain features, and create detailed geospatial data, supporting a wide range of applications, including urban planning, infrastructure development, and environmental management.

5. Navigation satellites play a critical role in search and rescue operations. Distress beacons equipped with satellite communication capabilities can transmit emergency signals received by satellites and forwarded to rescue coordination centres. These signals help locate individuals or vessels in distress, enabling swift and effective rescue efforts.

As technology advances, satellite-based navigation systems evolve to provide even greater accuracy, reliability, and coverage. Future developments include integrating multiple navigation systems, improved signal interference resistance, and enhanced indoor navigation capabilities.

Scientific Satellites

Scientific satellites encompass a wide range of missions, each designed to investigate specific phenomena or gather data in different regions of the electromagnetic spectrum. These satellites are equipped with sophisticated instruments that enable detailed observations and measurements, providing valuable insights into the nature of the Universe.

Scientific satellites encompass a wide array of missions designed to investigate various scientific phenomena. These missions often focus on specific areas of study, such as Earth observation, climate change, space weather, and fundamental physics. Let's explore a few notable examples:

1. Earth observation satellites enable scientists to study our planet from space, monitoring changes in land cover, tracking atmospheric conditions, and observing the dynamics of oceans and ice caps. These satellites provide essential data for climate studies, weather forecasting, disaster management, and environmental monitoring.

2. Besides the new James Webb Space Telescope and the famous Hubble Space Telescope, several space-based telescopes have contributed significantly

to our knowledge of the Universe. Examples include the Chandra X-ray Observatory, which observes X-ray emissions from cosmic sources, and the Spitzer Space Telescope, which studies the infrared Universe, providing insights into the formation of stars, planets, and galaxies.

3. Solar satellites are dedicated to studying the Sun and its effects on Earth's space environment. These satellites observe solar activity, such as solar flares, coronal mass ejections, and the solar wind. They provide crucial data for understanding space weather, which has implications for satellite communications, power grids, and astronaut safety.

4. Some scientific satellites carry out experiments to investigate fundamental physics questions. For instance, the Gravity Probe B mission aimed to test Einstein's theory of general relativity by precisely measuring the effects of Earth's gravity on orbiting gyroscopes. Such missions push the boundaries of our understanding of the laws that govern the Universe.

These scientific and astronomical satellites are not limited to individual missions but often collaborate with ground-based observatories and other space-based instruments. Data collected from these satellites are shared among scientists worldwide, fostering collaboration and enabling comprehensive studies of cosmic phenomena.

Deep Space Probes

In the vast reaches of space, human ingenuity has propelled a fleet of probes that venture where humans cannot, unravelling the mysteries of our cosmic neighbourhood and beyond. This is not an exhaustive list, nor a very detailed one, but even though some are speeding away from us right now, and others have long since completed their missions, they are an important category of space objects created by humankind.

Voyager 1 and *2* (1977): Launched in 1977, the Voyager probes embarked on an odyssey to explore the outer planets and beyond. Their daring flybys of Jupiter, Saturn, Uranus, and Neptune revealed unprecedented details about these distant worlds. They provided iconic images of the Great Red Spot on Jupiter and the rings of Saturn. Voyager 1, in particular, ventured beyond our solar system and into interstellar space, carrying the famous Golden Record, a time capsule of humanity's sounds and images.

Cassini-Huygens (1997): A joint NASA-ESA mission, Cassini-Huygens embarked on a mission to study Saturn and its moons. Cassini provided unparalleled insights into the ringed giant, capturing breathtaking images and discovering new moons. Meanwhile, Huygens descended to the surface of

Titan, Saturn's largest Moon, revealing a world with lakes and rivers of liquid methane and ethane.

New Horizons (2006): Sent on a daring mission to the outer reaches of our solar system, New Horizons provided the first close-up images of Pluto in 2015, challenging our perceptions of the dwarf planet. Continuing its journey into the Kuiper Belt, this probe offers insights into our cosmic neighbourhood's distant and icy realms.

Hayabusa 1 and *2* (2014): Japan's Hayabusa missions targeted asteroids. Hayabusa 1 collected samples from the asteroid's surface and returned them to Earth in late 2020, providing insights into the formation of the solar system and the origins of water and organic molecules.

Osiris-Rex (2016): Launched by NASA, Osiris-Rex is on a mission to study the near-Earth asteroid Bennu. Arriving in 2018, it conducted detailed observations and sampled the asteroid's surface in 2020. The samples returned to Earth in 2023 and might hold clues about the early solar system and the building blocks of life.

The *Double Asteroid Redirection Test* (DART) 2021 mission, led by NASA, was designed to test our ability to alter the motion of an asteroid in space. Its target was the binary asteroid system Didymos, consisting of a larger primary body and a smaller moonlet named Dimorphos. DART's mission objective deliberately crashed into Dimorphos, altering its orbit around Didymos. The data collected from this mission will contribute to our understanding of asteroid dynamics and bolster humanity's preparedness for potential future encounters with near-Earth objects.

Did you know that some of the space objects we launched stopped working? It is such a waste of space...

CHAPTER FOUR

The Debris Swarm

Space debris originates in many sources, reflecting the cumulative effect of decades of human activities in space. Each space mission, satellite launch, and orbital operation contributes to the growing population of debris fragments, creating a complex and interconnected network of potential hazards. It's a bit like driving and leaving behind litter, with the critical distinction that the space litter keeps flying at high speeds and can hit you or others over the head.

Rocket stages, part of the large propulsion systems that deliver payloads into orbit, significantly contribute to space debris. After completing their job, these stages often remain in orbit as abandoned objects. Over time, they can break apart due to natural decay or explosive venting of residual propellants, generating fragments that add to the debris population.

Satellite failures are another source of space debris. When a satellite malfunctions or reaches the end of its operational life, it can become a dormant object in space. These defunct satellites can remain in orbit for years, posing a risk of collision with operational spacecraft or generating debris if they break apart.

Occasionally, satellites or rocket stages are intentionally destroyed to prevent them from becoming a long-term threat. This intentional destruction is often done through controlled reentry or by moving the object into a decaying orbit where it will burn up upon reentry into Earth's atmosphere. However, even intentional destruction can contribute to producing smaller debris fragments that remain in orbit for extended periods.

One of the most concerning sources of space debris is collisions between satellites or with other debris fragments. In 2009, a catastrophic collision occurred between an inactive Russian satellite, *Cosmos 2251*, and an operational U.S. satellite, *Iridium 33*. This collision generated thousands of trackable debris fragments, further exacerbating the space debris problem. Such collisions can occur due to the increasing congestion of satellites and debris in certain orbits, especially at popular altitude bands like Low Earth Orbit.

Human-made space debris also results from anti-satellite weapon (ASAT) tests. United States, Russia, China, and India have successfully destroyed satellites in orbit, resulting in thousands of new fragments and further increasing the danger in some orbits. Fragmentation events, whether resulting from collisions, intentional destruction, or other factors, significantly contribute to the growth of space debris. When objects collide or experience high-energy impacts, they can shatter into numerous smaller pieces. These fragments can remain in orbit for extended periods, posing threats to operational spacecraft and propagating the risk of further collisions.

Space debris is not solely created by human activities. Micrometeoroids, tiny particles of dust and debris from outer space, also contribute to the space debris population. While individual micrometeoroids pose minimal risks, the cumulative effect of their constant bombardment can degrade the surfaces of spacecraft and satellites over time.

It is important to note that space debris exists at various altitudes and orbital planes, creating a complex and interconnected web of objects.

This vast and diverse debris population poses challenges for space traffic management as monitoring and predicting potential collisions becomes increasingly complex.

The Kessler Syndrome

The Kessler Syndrome, named after NASA scientist Donald J. Kessler, is a theoretical scenario that describes the potential chain reaction of collisions in space leading to the proliferation of space debris. The Kessler Syndrome postulates that as the density of space debris increases, the likelihood of collisions between objects also increases, resulting in a cascade effect that generates even more debris. This cycle of collisions and debris creation poses a significant threat to operational satellites and the long-term sustainability of space activities.

The concept behind the Kessler Syndrome is based on the principle of the conservation of momentum. When two objects collide in space, their momentum is transferred to each other, potentially altering their trajectories. If the collision is forceful enough, it can shatter the objects into numerous fragments, each becoming an independent piece of debris. These fragments can then collide with other objects, creating a domino effect that perpetuates the production of more debris.

The potential consequences of the Kessler Syndrome are far-reaching. As the density of debris increases, the probability of collisions rises exponentially, leading to more fragmentations and the creation of smaller debris pieces. This growing debris population poses a grave risk to operational satellites, crewed spacecraft, and future missions. Even small fragments travelling at high velocities in space can cause significant damage upon impact with satellites or other spacecraft, rendering them inoperable or destroying critical components.

One of the most concerning aspects of the Kessler Syndrome is the creation of debris clouds in specific orbital regions. These clouds, often called debris fields or debris belts, can concentrate a high density of debris fragments, making these regions exceptionally hazardous. Satellites or spacecraft that pass through these regions face an increased risk of collision, which can lead to catastrophic consequences.

Mitigating the Kessler Syndrome requires a comprehensive space traffic management and debris mitigation approach. Efforts are focused on reducing the creation of new debris through responsible space practices, such as post-mission disposal or deorbiting satellites at the end of their operational life. Additionally, active debris removal (ADR) missions are being explored to capture and remove existing debris from orbit, reducing the overall density of debris and mitigating collision risks.

International collaboration is essential to prevent the Kessler Syndrome from becoming a reality. Space agencies, governments, and organizations work together to develop guidelines, policies, and best practices for debris mitigation and space traffic management. Improved tracking and surveillance systems, coupled with enhanced space situational awareness, are also crucial for the early detection of potential collisions and effective spacecraft manoeuvring to avoid debris.

The Kessler Syndrome is a stark reminder of the fragility of space infrastructure and the need for responsible space activities. By actively addressing the issue of space debris and implementing measures to prevent its proliferation, we can safeguard the orbital environment and ensure the continued exploration and utilization of space for future generations.

Are we there yet? No, but the system is rerouting the policymaking due to traffic.

Satellite Vulnerability

Satellites, the critical assets orbiting Earth, are highly vulnerable to the threats posed by space debris. As the space debris population grows, the risk of collisions with operational satellites increases, potentially leading to disastrous consequences.

Each additional debris fragment adds to the complexity of satellite operations, requiring increased vigilance, regular tracking, and constant monitoring to ensure their safety.

Understanding the satellites' vulnerabilities is essential for developing strategies to protect them and ensure the continuity of their missions.

As they orbit the Earth at high velocities, typically from several kilometres per second, when a satellite collides with space debris at such high speeds, even small debris fragments can have a devastating impact. The kinetic energy released upon collision can cause severe damage to the satellite's structure, critical components, solar panels, antennas, and delicate instruments.

Satellites are not designed to withstand direct impacts from space debris. Their structures are typically lightweight and optimized for efficient operation in space. While some satellites incorporate shielding measures, such as Whipple shields or bumper-like structures, they provide limited protection against high-velocity debris impacts. The sheer number of debris fragments and their unpredictable trajectories make it nearly impossible to physically shield satellites from every potential collision.

Even if a satellite does not experience a direct impact, it is still susceptible to surface damage caused by micrometeoroids and smaller debris particles. Over time, the cumulative effect of these impacts can degrade the satellite's outer layers, solar panels, and thermal protection systems.

Surface damage can compromise the satellite's functionality, degrade its performance, and increase the risk of subsequent collisions.

In addition to the threat of direct impacts, satellites face the risk of fragmentation when a nearby object, such as a defunct satellite or rocket stage, experiences a collision. The resulting fragments can propagate and spread throughout orbital paths, increasing the density of debris and heightening the risk of further collisions. These cascading events can create additional hazards for satellites as they navigate increasingly cluttered orbits.

Satellites often have limited propulsion capabilities, making it challenging to perform evasive manoeuvres to avoid potential collisions. Maneuvering requires careful planning, fuel consumption considerations, and precise orbital calculations. In scenarios where the risk of collision is high and the reaction time is limited, satellites may not have sufficient resources or time to execute avoidance manoeuvres effectively.

Space agencies and satellite operators implement various strategies to address the vulnerability of satellites. These include active monitoring of space debris, collision avoidance manoeuvres when necessary, and designing satellites with enhanced robustness to withstand potential impacts. Additionally, ongoing research and development efforts focus on advanced shielding technologies, such as lightweight materials with improved resistance to debris impacts.

Risks to Human Spaceflight

Human spaceflight represents the pinnacle of human exploration and achievement, but it also comes with inherent risks, including those posed by space debris. Protecting astronauts and their spacecraft from the hazards of space debris is crucial to ensuring their safety and the success of human space missions.

The primary risk that space debris poses to human spaceflight is the potential for collisions with crewed spacecraft. A collision with a piece of debris could puncture the spacecraft's hull, damage critical systems, or compromise life support systems. The safety of astronauts is paramount, and avoiding collisions with debris is a top priority for mission planners and operators.

Crewed spacecraft, such as space stations or future deep-space exploration vehicles, have larger surface areas than satellites. This increased size makes them more susceptible to debris impacts. Additionally, crewed vehicles have more complex systems and delicate instruments, making even minor damage critical. The safety of astronauts and the integrity of their spacecraft rely on effective shielding and debris avoidance measures.

Moreover, human space missions often involve extended durations in space, whether it is a stay on a space station or long-duration exploration missions to the Moon or Mars. The longer astronauts are exposed to the space environment, the greater the cumulative risk of encountering debris. Over time, the probability of a collision increases, necessitating enhanced vigilance and monitoring to ensure the crew's safety.

Crewed spacecraft cannot simply be abandoned in the event of an impending collision, and there are limited options for escape or evacuation. The space around might be ample, but it is harsh. There are no parachutes, no shelters, just the void. This limitation increases the importance of reliable debris tracking and space situational awareness to provide timely warnings and allow for necessary manoeuvring or sheltering inside the spacecraft.

In addition to the immediate dangers of physical damage from debris impacts, space debris also poses health risks to astronauts. The impact of debris on the spacecraft can generate secondary fragments or release hazardous substances, such as toxic propellants or materials. These fragments or substances could enter the crew compartment, potentially endangering the health and well-being of the astronauts.

Mitigating the risks posed by space debris to human spaceflight involves a multi-faceted approach. This includes prevention through comprehensive space traffic management, accurate tracking and cataloguing of debris, and precise prediction of collision risks. Damage minimization is achieved through spacecraft shielding and appropriate design considerations.

What goes up generally comes down...

Down-to-Earth Consequences

Space debris colliding with a communication satellite can disrupt telecommunications networks, interrupting phone, internet, and broadcasting services for vast areas of Earth. Similarly, damage to weather satellites can hinder accurate weather forecasting and monitoring, impacting disaster response and resource management.

Certain space-based systems are of utmost importance for national security and defence operations. These include reconnaissance and surveillance satellites, early warning systems, and satellite-based navigation systems. Any disruption to these critical infrastructure elements due to debris collisions could have severe consequences, compromising the ability to monitor threats, respond to emergencies, or conduct military operations effectively.

When space debris hurtles toward Earth, it undergoes a fiery journey through the atmosphere, facing intense heat and friction. Most smaller debris, often no larger than a grain of sand, disintegrates during this descent, creating fleeting streaks of light known as meteors. However, larger fragments pose a more complex challenge. While some may burn up entirely, others may survive the fiery plunge, reaching the Earth's surface. That does not mean that every meteor we see is a space debris. The Universe tends to throw a lot of rocks at us, and sometimes they are pretty big.

According to NASA, an average of one catalogued piece of debris has fallen back to Earth each day for the past 50 years. Despite their size, there has been no significant property damage from the debris.

L.W, a resident of Tulsa in Oklahoma, US, is the first person recorded as being struck by space debris, in 1997. The debris was about the size of a hand, did not cause much harm and is thought to have come from an American Delta II rocket. Another incident that involved humans being harmed was recorded in 1969 when five sailors on a Japanese ship were injured when space debris from what was believed to be a Soviet spacecraft struck the deck of their boat.

As damages kept being recorded, the first fines appeared. In 1978, Canada sued Russia over debris that rained down on its territory from a nuclear-powered satellite, eventually settling on an amount of more than $2 million.

In 1979, the Australian town of Esperance jokingly fined NASA $400 after pieces of its Skylab space station fell in the region.

In 2018, the U.S. startup Swarm Technologies was fined $900,000 by the Federal Communications

Commission (FCC) for launching satellites without permission.

The year 2023 marks a first as the FCC in the United States issued its first fine for space debris still in orbit, ordering the US TV provider Dish to pay $150,000 for failing to move one of its satellites into a safe orbit, thus violating rules against space debris proliferation. The EchoStar-7 satellite should have deorbited in May 2022, when it ran out of fuel, forcing the Dish corporation to abandon it 100 miles (178 km) short of its intended disposal location far above geostationary orbit.

Damages are not only the result of direct impacts. Light pollution from space objects can interfere with astronomical observations, affecting both amateur stargazers and professional astronomers. The reflection of sunlight off satellite surfaces creates streaks of light across the night sky, disrupting the natural darkness crucial for celestial observations. In photography-based observations, these new moving lights impair the capacity to detect other movements, like asteroids and comets.

Additionally, the radio signals emitted by satellites can contribute to radio frequency interference, posing challenges for terrestrial communication systems and radio astronomers. It is like trying to listen to a whisper in a room full of crickets. I think that aliens will need to yell pretty loud in these conditions. And carry a big light.

Last but not least, the launches and reentries add particles to the atmosphere, thus contributing to changes that could have adverse effects. As precise measurements are complex to make, this phenomenon requires further studies, but what is certain is that some materials used in space object construction or propulsion are hazardous to human health and biodiversity.

As humanity sets its sights on ambitious space exploration goals, including crewed missions to the Moon, Mars, and beyond, the threat of space debris becomes even more significant. Debris in certain orbital regions can hinder the safe passage of spacecraft, increase collision risks, and limit the available operational space for future missions. To help make sense of the traffic, scientists have devised ways of identifying where the best spots for satellites are and how to navigate safely.

Figure 1 Consequences of space activities

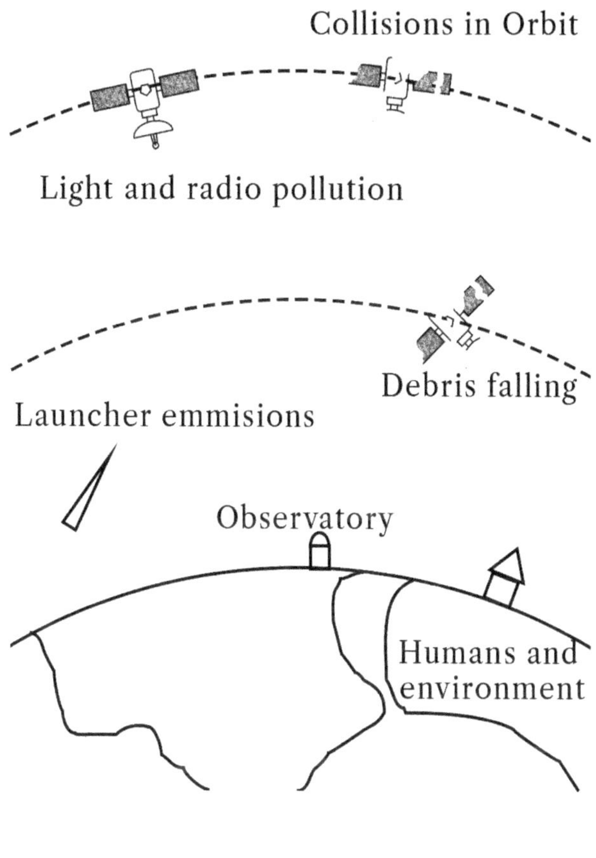

Not to scale

CHAPTER FIVE

Celestial Highways

Until this point in the book, you have a short history of humankind's capacity to launch useful things into space. Some people also launched less useful stuff there, like a car, for example, but all these have some common points. The ones we will explore further are physics and human-made acronyms.

The designations of Low Earth Orbit (LEO), Medium Earth Orbit (MEO), and Geostationary Earth Orbit (GEO) serve to categorise and differentiate orbits based on their specific characteristics and applications. These distinctions are essential for efficient space operations and satellite deployment.

Imagine you're in a spaceship and want to stay in space without falling back to Earth. Earth has a lot of mass, so it has a strong gravitational pull.

But here's the exciting part: when you're in a spaceship going really fast, you can balance that gravitational pull and stay in orbit. It's like when you throw a ball. If you throw it too slowly, it quickly falls back to the ground. If you throw it really fast, it goes further away before falling back. So, if you throw a thing with more power, it will go very far. Enough force and the curved line that the object follows becomes an orbit.

In space, it's a constant tug-of-war between your spaceship wanting to go straight and gravity trying to pull it down. If you get the speed just right, these forces balance out, and you end up in a stable orbit. The closer you are to Earth, the faster you need to go to stay in orbit.

There are different kinds of orbits too. Imagine swinging a yo-yo around you. If you do it slowly, the yo-yo comes close, and if you do it fast, the yo-yo swings out wider. Orbits work the same way. Low Earth Orbit is like swinging the yo-yo close, Medium Earth Orbit is a bit further, and Geostationary Earth Orbit is like swinging it far out.

So, when you see satellites up there or astronauts on the International Space Station, they're not just floating – they're actually falling towards Earth, but their sideways speed is so fast that they keep missing the ground. It's like a perpetual game of cosmic tag, with gravity always chasing but never catching.

And now that you have a mental picture of an orbit, let's take a small step towards physics and orbital mechanics. One does not need to remember all these right now. Still, when you encounter references later, it is handy to remember that the explanation is here.

Several key elements define an orbit and determine the characteristics of the path that an object takes as it moves around another massive object, such as a planet or a star.

These orbital elements are collectively known as the Keplerian elements, named after Johannes Kepler, who first described the laws governing planetary motion. Together, these elements comprehensively describe an orbit's size, shape, and orientation. In practical terms, these elements are crucial for spacecraft and satellite design, navigation, and mission planning, as they allow scientists and engineers to predict the trajectory of objects in space.

For those curious about the specifics, without going into the science behind them, the elements describing an orbit are: semi-major axis (a), eccentricity (e), inclination (i), longitude of the ascending node (capital omega), argument of periapsis (omega), mean anomaly (v)

Figure 2 Keplerian elements

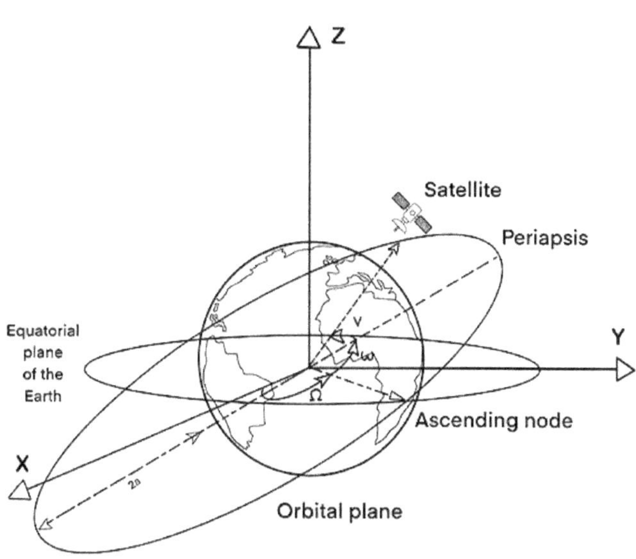

Not to scale

Celestial objects, like planets or satellites, can follow various paths around a central body, such as a star or a planet. Circular and elliptical orbits are two common types of paths.

In a circular orbit, the object travels in a perfect circle around the central body. The distance between the object and the centre remains constant throughout the orbit. Circular orbits are characterised by a single, fixed radius, providing a consistent and symmetrical path. Like Earth around the Sun, planets in our solar system have nearly circular orbits. Nearly.

On the other hand, in an elliptical orbit, the object follows an oval-shaped path around the central body. The distance between the object and the centre varies as it moves along the elliptical path. Elliptical orbits are described by the semi-major axis (the longer radius of the ellipse) and the eccentricity (a measure of how stretched or circular the ellipse is). Comets and many satellites often have elliptical orbits.

In summary, the main difference lies in the path's shape: circles are symmetrical and constant in radius, while ellipses are elongated and have varying distances from the centre. The choice between circular and elliptical orbits depends on a mission's specific requirements or celestial bodies' natural dynamics.

Congratulations on making it here, and as a reward, here is a simple graphical representation of the orbits using the abovementioned notions.

Figure 3 Shapes of Orbits

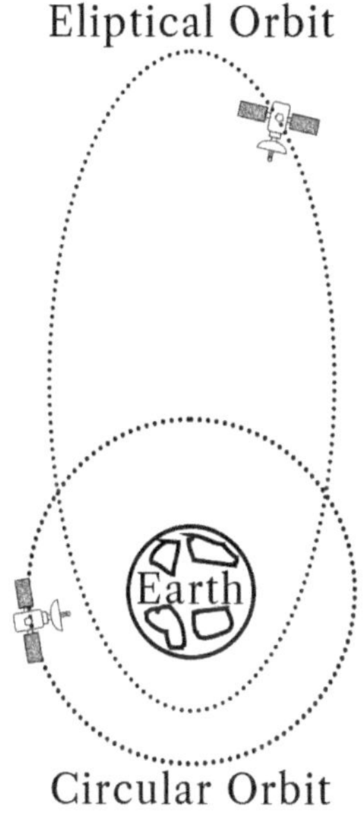

Not to scale

Navigating the Layers of Space

If the definitions above have not brought tears to your eyes, let's think about an onion for a moment. In our attempt to create a human order in a Universe that cares little about anything else but the laws of physics, humans have taken the example of the onion and named some parts of the space enveloping Earth.

The layered designations of Low Earth Orbit (LEO), Medium Earth Orbit (MEO), and Geostationary Earth Orbit (GEO) serve to categorise and differentiate orbits based on their specific characteristics and applications. And yes, scientists did not use the word orbit in two ways to confuse you. In the end, one object can follow an elliptic path (elliptical orbit) in space and be at its closest point to Earth in Low Earth Orbit and at its farthest, outside the Solar system.

LEO is a region of space situated relatively close to Earth's surface, extending from approximately 180 to 2,000 kilometres above sea level. In this orbital zone, satellites and spacecraft circle the planet in a near-weightless environment, experiencing minimal atmospheric drag. Do you remember the orange analogy from Chapter 3?

LEO's height is defined in several ways. Throughout an elliptical orbit, the height of an object can change dramatically. The Earth is not a perfect ball and is flatter at the poles than at the equator.

This is called oblateness. Due to the oblateness of Earth's spheroid form and local topography (mountains, oceans), even for circular orbits, the height above ground might vary by as much as 30 km, particularly for polar orbits. This is why definitions based on altitude are inherently vague.

That said, most LEO region orbits are included in a specific range. That range is defined by an orbit period of 128 minutes, corresponding to a semi-major axis of 8,413 kilometres (5,228 miles). This is why the upper altitude restrictions, in some definitions of LEO, are set at 2,042 kilometres (1,269 miles) beyond the average radius of Earth for circular orbits.

LEO is a primary location for various space activities, including satellite deployment, scientific research, and human spaceflight missions. Space objects in LEO have relatively short orbital periods, often completing an orbit around the Earth in 90 minutes or less. This proximity allows for Earth observation, reconnaissance, and rapid data collection but requires more frequent orbital adjustments due to atmospheric drag.

Satellites in LEO can provide crucial functions such as Earth observation, telecommunications, and navigation. The proximity to Earth allows for shorter communication delays and facilitates relatively accessible launch and rendezvous capabilities.

Notable structures within LEO in 2024 include the International Space Station and the Tiangong space station. So far, there have been no inhabited space stations beyond LEO.

Since so many satellites are in LEO, any objects whose orbit passes through this region must be closely monitored, even if they travel below or above it.

Figure 4 The Orbital Layers

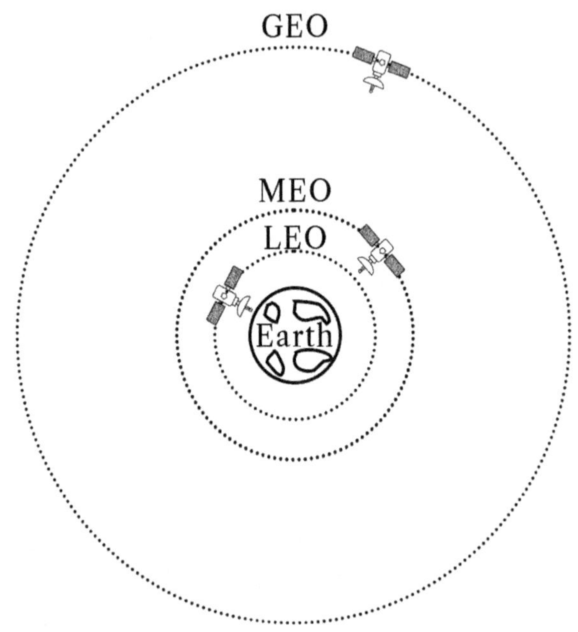

Not to scale

Medium Earth Orbit (MEO) is a region of space situated above LEO and below Geostationary Earth Orbit (GEO), typically ranging from around 2,000 to 35,786 kilometres above Earth's surface. Space objects placed in MEO occupy a sweet spot in terms of orbital altitude, offering some of the advantages of both LEO and GEO. Satellites in MEO experience less atmospheric drag than in LEO, allowing for longer operational lifespans. The moderate orbital periods of MEO satellites strike a practical compromise, providing a good balance between coverage and revisit times for global navigation and communication purposes.

A particular case is satellite constellations, where a satellite in semi-synchronous orbit at an altitude of roughly 20,200 kilometres has a 12-hour orbital period and passes over the same two equatorial points daily. The Global Positioning System (GPS) constellation uses this very predictable orbit. Other navigation satellite systems using it are the Russian GLONASS (at an altitude of 19,100 kilometres), the European Union Galileo (at an altitude of 23,222 kilometres), and the Chinese BeiDou (at an altitude of 21,528 kilometres), all using comparable MEOs.

The second common MEO is the Molniya orbit. Molniya, or Lightning in Russian, is named after Molniya satellites, a series of Soviet/Russian civilian and military communications satellites which have used this type of orbit since the mid-1960s. A geostationary orbit is valuable for the constant view it provides, but satellites in a geostationary orbit are parked over the equator, so they don't work well for far northern or southern locations, which are always on the edge of view for a geostationary satellite. The advantage of the highly elliptical Molniya orbit with an apogee over Russian territory is that it works well for observing high latitudes.

Geostationary Orbit (GEO) is a specialised region in space located approximately 35,786 kilometres above the Earth's equator, where satellites orbit the planet at the same rotational speed as the Earth. This synchronised motion allows these satellites to remain stationary relative to a fixed point on the Earth's surface, creating the appearance of a stationary point in the sky. Satellites in the GEO complete one orbit around the Earth in precisely 24 hours, aligning with the Earth's rotation period. This unique characteristic makes GEO ideal for specific applications, such as communication and weather observation, where a consistent and unchanging vantage point is crucial. As a result, many communication satellites, television broadcasting satellites, and weather satellites are strategically positioned in the Geostationary Orbit, ensuring continuous and reliable coverage over specific regions on the Earth's surface.

And if this was not enough, and we needed more names designating the imaginary layers of space, the region that contains the GEO as a particular case is called High Earth Orbit (HEO). The acronym should not be mistaken for Highly Elliptical Orbit - HEO, which does not refer to the layer but to the orbit's shape. Is it clear, or is it that the peeling of the onion has left us in tears?

Lagrange points

The designations of the layers are essential for knowing how far from Earth a space object is, thus allowing traffic coordination by putting objects in particular regions based on their mission. But in addition to layers, some specific points in space are essential for space missions due to their physical characteristics.

Isaac Newton formulated the laws of mechanics and gravitation at the end of the 17th Century. French mathematician Henri Poincaré described the Three-Body problem in the same period. This problem involves predicting the motions of three celestial bodies under the influence of their mutual gravitational forces. While the motions of two bodies can be precisely calculated, introducing a third introduces a level of complexity that defies straightforward solutions.

The essence of the Three-Body Problem lies in the inherent instability and sensitivity to initial conditions. Tiny variations in starting positions or velocities can lead to vastly different outcomes over time, making long-term predictions challenging. The mathematical complexities of solving the Three-Body Problem have inspired countless researchers and led to the development of chaos theory.

Despite its elusive nature, the Three-Body Problem has real-world applications in astrophysics, space exploration, and celestial mechanics. Understanding the gravitational dynamics at play helps scientists model celestial bodies' trajectories, predict planetary systems' behaviour, and design space missions with precision.

This theory directly influenced the discovery and use of Lagrange points. Named after the Italian mathematician, physicist and astronomer, later naturalised French, Joseph-Louis Lagrange (1736-1813), these five locations exist in specific relationships between two massive bodies, such as a planet and its moon or a planet and the Sun, where gravitational forces create delicate points of equilibrium.

The Lagrange points, denoted L1 through L5, represent positions where the gravitational forces of the two massive bodies balance, allowing smaller objects to remain in a stable position relative to them. L1 is located along the line connecting the two massive bodies, while L2 is directly opposite, forming a straight line with them. L3 also lies along this line but on the opposite side of the secondary massive body. Lagrange points L4 and L5, often called the Trojan points, form equilateral triangles with the two massive bodies, creating stable regions around 60 degrees ahead and behind the secondary massive body in its orbit.

These Lagrange points serve as celestial parking spots for satellites and space probes, enabling them to maintain a fixed position relative to Earth or other celestial bodies.

In particular, the L1 and L2 Lagrange points are popular destinations for space missions. The first was in 1978 with the International Sun-Earth Explorer (ISEE) programme, whose modules were placed in orbit around the L1 Lagrange point of the Sun-Earth system (located at 1.5 million km, that is, at less than 1% of the Earth-Sun distance). In 2018, the L1 point in the Earth-Moon system (326,000 km from Earth, 15% of the Earth-Moon distance) also became the home to the Chinese relay satellite Queqiao, which communicates with the Chang'e 4 lunar probe on the far side of the Moon.

The L2 point of the Sun-Earth system, also located about 1.5 million km from the Earth but in the opposite direction to L1, is home to the 2009 Planck satellite, the 2015 LISA Pathfinder satellite, and the Gaia mission. The latest satellite to reach L2 is the James Webb Space Telescope3, launched on 25 December 2021. Since objects in these points have definite advantages in terms of physics, like in the case of orbits, good management of what is there is vital.

Figure 5 Lagrange points of the Sun-Earth system

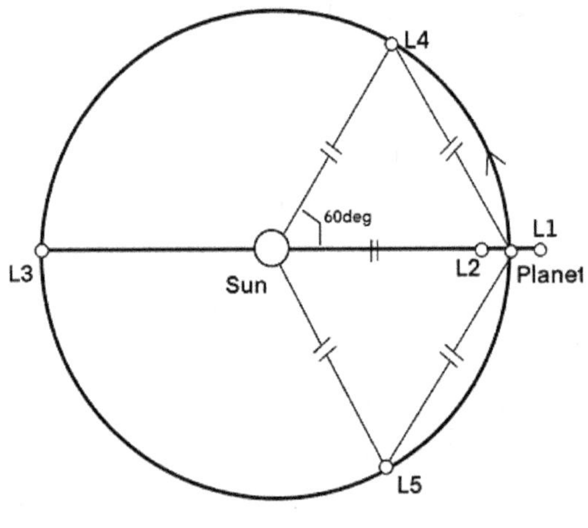

Not to scale

CHAPTER SIX

Bring order to chaos

The last few decades have witnessed an unprecedented growth in the number of satellites orbiting the Earth. What was once a realm dominated by a handful of government-operated spacecraft has now transformed into a bustling domain with a diverse array of satellites from government agencies, commercial entities, and even individual researchers.

Commercialisation of space is the main factor in the rapid expansion of the satellite population. With the rise of commercial space companies, satellite services such as telecommunications, Earth observation, and global positioning systems have become more accessible and commercially viable.

The lower prices, increased access and growing demand for these services have fueled the deployment of an increasing number of satellites.

One of the most prominent trends driving the surge in satellite numbers is the emergence of space-based internet constellations. Companies like SpaceX, OneWeb, and Amazon are developing massive constellations comprising thousands of interconnected satellites in LEO. The ambitious scale of these projects has contributed significantly to the rapid growth of the satellite population as multiple companies strive to deploy their own constellations.

Another driver of the satellite population boom is the advent of small satellites, often called CubeSats or nanosatellites. Such miniature spacecraft, typically weighing less than 10 kilograms, have opened access to space for a broader range of organisations and individuals. Their compact size and lower costs make them attractive for universities, startups, and research institutions conducting space-based experiments and technology demonstrations.

Government agencies and national space programs continue to launch satellites for various purposes, including scientific research, Earth observation, weather monitoring, and national security. Spacefaring nations are keen to maintain their presence in space and leverage satellite technology for a wide range of applications. As new countries enter the space race and established players expand their programs, the number of satellites deployed by national entities continues to rise.

To tackle the challenges posed by the increasing satellite population, space agencies, international organisations, and industry stakeholders are actively working on various initiatives. These include the development of standardised protocols for collision avoidance and manoeuvre planning, establishing space traffic coordination centres, and enhancing space situational awareness capabilities. Additionally, efforts are being made to promote responsible space practices, such as implementing end-of-life disposal plans for satellites to minimise the creation of space debris. However, a new form of land grabbing is now being observed, with radio frequencies, orbits and points in space being strategically occupied to ensure future access and deny competitors access.

The History of Space Traffic Management

As a current point of interest in space law and policy, numerous studies have been conducted on space traffic management (STM). However, there is still a great deal of ambiguity regarding STM's nature, scope, and goals, as each study tends to come up with its own description and definition. Many people are unfamiliar with it, so let's find out where the idea came from and how it has evolved over time.

A Czechoslovak attorney named Vladimir Mandl authored "Das Weltraum-Recht" in 1932 when space flight was still the stuff of science fiction and the dreams of creative authors and filmmakers. The paper considered similarities with air traffic protocols and is widely regarded as the first comprehensive work on space law. Mandl proposed the establishment of a new area of law to supplement existing air law restrictions in light of the anticipated legal ramifications stemming from the expansion of space activities. He introduced the concept of "space traffic rules" for the first time in his analysis of the legal difficulties associated with astronautics.

By the 1960s and 1970s, numerous works were published on space traffic management-related topics, with these two decades being pivotal and marking the start of space ventures. So, along with these leaps forward came a new policy. Eugène Pépin, in 1957, outlined five facets that would necessitate the development of "regulatory rule(s)" for circulation in outer space:

- The ascent of rockets through sovereign airspace;
- The re-entry of rocket bodies;
- Accidental collisions between orbiting satellites;

•The need for identification of satellites in case of an accident;

•The avoidance of harmful radio interference.

It is worth noting that a worldwide geophysical research program conducted from July 1957 to December 1958, the International Geophysical Year, used satellites. After analysing the data, most of the leading space experts voiced worries about electromagnetic interference and called for rules regarding such risks.

Fast forward to 1982, when astrophysicist Dr. Lubos Perek published the paper "Traffic Rules for Outer Space", providing the first concrete suggestion for a set of rules. These regulations were more in-depth than Pépin's five elements, although they still broadly correspond to them; the addition of space debris mitigation is most notable.

The necessity to establish and enforce traffic laws became practical on the 24th of July 1996, when the French military satellite Cerise was damaged by a piece of debris, making it the first casualty of a human-generated accident in outer space. So, by the late 1990s and beyond, conversations and collaborations regarding space traffic safety had become far more serious.

At the American Institute of Aeronautics and Astronautics (AIAA) 5th and 6th International Space

Cooperation workshops in 1999 and 2001, the idea of space traffic and the desire to control it made a clear and obvious resurgence. The International Academy of Astronautics (IAA) formed an STM working group that eventually published the 2006 IAA Cosmic Study on Space Traffic Management, which addressed orbital management, collision avoidance, relevant orbital debris issues, and regulatory framework needs.

The researchers were already starting to characterise STM as the interplay between an STM system—the collection of technical capabilities that ensure the security and longevity of space—and an STM regime—the framework of laws, policies, and regulations that ensure these things are implemented.

Managing space traffic is an iterative procedure. Technological advancements and changes in the type of both debris and clean-up methods make the field everchanging. Measures must be taken to ensure the continued viability of space operations and the protection of the space environment.

While these goals are what most people think of when they hear the term "space traffic management," the term has been defined in various ways throughout the years, leading to some confusion.

Arguably, there may be no single definition of STM in any definitive or universally accepted sense.

That said, several papers have proposed their own interpretations of the term, frequently placing it within the context of their own goals, arguments, or concerns. Tracking and an improved understanding of the probabilities of collisions at specific locations and altitudes will be an aspect that will always need further investigation.

The European Space Agency believes there are about 36,000 bits of space junk greater than ten centimetres (10 cm or 3,94 inches is the size of a regular orange). The estimated number of particles with diameters ranging from 1 to 10 cm is 500,000. The quantity of particles bigger than one millimetre reaches one hundred million. The amount of material circling the Earth had surpassed 9,000 metric tons as of January 2022.

The U.S. Space Surveillance Network frequently tracks large orbital debris (> 10 cm or 3,94 inches). Ground-based radars can detect objects as tiny as 3 mm, allowing for a statistical approximation of their quantities. Examining impact characteristics on the surfaces of returning spacecraft can provide estimates of the amount of orbital debris shorter than 1 mm, albeit this has been confined to spacecraft operating at altitudes less than 600 km.

Before 2007, the most common cause of debris was explosions of the upper stages of launch vehicles and spacecraft. Several unrelated incidents, like the intentional destruction of China's Fengyun-1C weather satellite in 2007 and the unintentional collision of the American communications satellite Iridium-33 with the retired Russian spacecraft Cosmos-2251, changed that.

Most orbital debris is found in a region extending upwards of 2,000 km of Earth's surface. Incidentally, this region is included in LEO, where most human spaceflight activities take place. The total quantity of debris inside this volume varies substantially with altitude. The highest density of debris observed is between 750 and 1000 kilometres.

This could beg the question, how does the ISS avoid debris? Does it have to? Suppose another object is expected to pass within a few kilometres of the Space Station. In that case, it will generally manoeuvre away from the object if the likelihood of a collision is calculated to be 1 in 10,000. This happens just about once a year on a typical basis. If the space debris situation gets worse, how would it be managed?

The International Space Station is the most extensively insulated spaceship ever flown. Critical components, such as livable sections and high-pressure tanks, should survive the impact of debris up to 1 cm in diameter. The likelihood of a major ISS component being hit by debris over 1 cm

(0,39 inches) in size is considered low, and methods to mitigate this risk are being researched.

During the Mir space station orbits, there was also damage. Mir's exterior photographs indicate several hits from tiny space debris and even meteoroids. The massive, vulnerable solar panels that couldn't be shielded from tiny particles suffered the most damage. However, this did not spoil its mission.

The higher the elevation of space debris, the more it remains in Earth's orbit. Debris remaining in orbits less than 600 km altitude often falls back to Earth after a few years. At an altitude of 800 km, the duration for orbital degradation is commonly measured in hundreds of years. Orbital debris over 1,000 km will often continue to circle the Earth for over a millennium or more.

Space traffic management notions

While the fundamental principles have remained the same, the overall structure of the limits of what Space Traffic Management is has seen some variations. A 2016 report by the US-based Institute for Defense Analyses' Science and Technology Policy Institute (IDA-STPI) proposed to remove the technical aspects of the traffic issue, mostly Space Situational Awareness (SSA), from the scope of STM. IDA-STPI envisions STM primarily as a regime focusing on coordination and separating SSA and STM.

A 2020 European Space Policy Institute (ESPI) report suggested dividing STM into three complementary functions: space traffic monitoring, regulation, and coordination.

Conversely, The International Astronautical Federation's STM Terminology Working Group proposed differentiating between an STM regime focusing on safety issues and a regime dealing with long-term sustainability concerns. The latter would address debris mitigation and remediation, also known as space environment preservation.

Space traffic management begins with comprehensive monitoring and tracking of space objects, including active satellites, debris fragments, and rocket stages. Ground-based radar systems, telescopes, and other tracking technologies are used to continuously monitor these objects' positions, trajectories, and characteristics. When the risk of a collision is identified, operators can manoeuvre satellites to safe distances or adjust their orbits to avoid the debris. Effective collision avoidance requires accurate tracking data, timely information sharing, and coordination among satellite operators.

Space Situational Awareness is a complex system that provides knowledge and understanding about space hazards. It comprises:

· Space Surveillance and Tracking (SST): a system of sensors to survey and track space objects, with processing capabilities to provide data, information and services concerning objects that orbit the Earth.

· Near-Earth Objects (NEO): capabilities to monitor the risk of natural space objects approaching the Earth, such as asteroids and comets.

· Space Weather Events (SWE): capabilities to monitor space weather and solar activity.

Improved SSA enables better prediction and assessment of collision risks, allowing operators to take proactive measures to avoid potential collisions. It also facilitates early warning systems and enables timely response to any developing situations.

International space organisations and regulatory bodies have developed debris mitigation guidelines and best practices to minimise the creation of new debris. These guidelines outline measures that satellite operators and launch providers should follow to limit debris generation during launches and operations. They include post-mission disposal and deorbiting procedures to remove satellites from orbit at the end of their operational life.

Space traffic management is not confined to a single country or organisation. It requires international cooperation and the development of regulatory frameworks to address the global nature of space activities. International agreements, such as the United Nations Space Debris Mitigation Guidelines, provide a framework for responsible space practices and encourage information sharing, data exchange, and coordination among nations to ensure space's safe and sustainable use.

Technological advancements play a crucial role in improving space traffic management capabilities. Improved tracking systems, including radar and optical sensors, allow for more precise tracking of space objects. Artificial intelligence and machine learning algorithms can help process vast amounts of tracking data, enabling better predictions and assessments of collision risks. Additionally, advancements in propulsion systems and autonomous satellite operations contribute to enhanced manoeuvrability and collision avoidance capabilities.

Raising public awareness about the challenges posed by space debris is an essential aspect of space traffic management. Educating the public about the risks and consequences of space debris fosters a sense of responsibility and promotes support for initiatives aimed at mitigating debris and preserving the space environment.

By the time you finish this book, you will also be ready to share the knowledge! Remember to leave a kind review.

Increased public awareness can encourage governments and space agencies to allocate resources and support research and development efforts related to space traffic management.

Space Traffic Coordination

Without proper coordination, the risk of collisions and close approaches increases, potentially leading to the loss of valuable assets and the creation of more space debris. One thing is to be hit by a random object, and another for two controlled crafts to collide.

Space traffic coordination is essential to ensure satellites' safe and responsible operation and protect critical orbits for future use.

Several international organisations play a significant role in space traffic coordination, fostering collaboration among spacefaring nations and promoting responsible space activities:

1. United Nations Office for Outer Space Affairs (UNOOSA): UNOOSA plays a central role in promoting international cooperation in space

activities. It facilitates the exchange of information, supports the development of space policies and guidelines, and encourages responsible space practices among member states.

2. International Telecommunication Union (ITU): The ITU is responsible for coordinating and allocating radiofrequency spectrum and orbital slots for satellite communication. It plays a crucial role in minimising interference between satellite systems and ensuring efficient use of the limited frequency spectrum.

3. International Astronautical Federation (IAF): The IAF brings together space agencies, industry representatives, and academic institutions to promote cooperation and exchange knowledge in space activities. It organises conferences, workshops, and forums to address various space traffic coordination and management aspects.

4. Committee on Space Research (COSPAR): COSPAR focuses on scientific research and exploration in space. It promotes interdisciplinary collaboration and addresses issues related to space traffic coordination, space debris mitigation, and planetary protection.

Dedicated centres have been established to monitor, track, and manage space activities to enhance space traffic coordination. These centres provide valuable services such as conjunction analysis, collision warnings, and manoeuvre planning assistance.

Some notable centres include:

1. United States Space Surveillance Network (SSN): Operated by the United States Space Command, the SSN is responsible for tracking and cataloguing space objects, detecting potential collisions, and providing data to satellite operators for collision avoidance manoeuvres.

2. European Space Operations Centre (ESOC): The ESOC, operated by the European Space Agency (ESA), is responsible for mission planning and satellite operations. It supports space traffic coordination by providing conjunction analysis and manoeuvre planning services.

3. Joint Space Operations Center (JSpOC): The JSpOC, operated by the United States Department of Defense, monitors and tracks space objects, provides space surveillance data, and collaborates with international partners to enhance space traffic coordination and safety.

As you see above, the entities mentioned are either U.N., U.S. or European. That does not mean that other countries are not participating or actively looking at the issue. Russia, China, Japan, and India have made their own steps in STM and SSA, sometimes independently and other times in cooperation with others.

In the European Union following the European Commission's decision from 3 June 2022, the European Union Agency for the Space Programme (EUSPA) has taken responsibility for the E.U. Space Programme's SST Front Desk operations starting from 1 July 2023. The responsibility was transferred from the European Union Satellite Centre (SatCen) to EUSPA's Galileo Security Monitoring Centre (GSMC) in Madrid.

This activity is done in cooperation with the Space Surveillance and Tracking Partnership that in 2024 includes: Austria (FFG), Czech Republic (MDCR), Denmark (RDAF), Finland (FMI), France (CNES), Germany (German Space Agency at DLR), Greece (NOA), Italy (ASI), Latvia (IZM), the Netherlands (EZK), Poland (POLSA), Portugal (PT MoD), Romania (ROSA), Spain (AEE) and Sweden (SNSA).

The Russian Federation has its own Hazard Alarm System (ASPOS) that provides predictions to operators and ROSCOSMOS, the State Space Corporation.

As space traffic coordination requires collaboration among satellite operators, space agencies, and industry stakeholders, various other initiatives, private or involving states, have been established to facilitate this collaboration.

Orbital Slot Allocation

In the realm of space traffic management, one of the key aspects is the allocation of orbital slots. Orbital slots refer to specific positions in space where satellites are assigned to operate. Think of them like traffic lanes and parking spots.

These slots are defined by their altitude, inclination, and longitude; each slot has unique characteristics and advantages. The allocation of orbital slots is crucial for maximising the efficiency and utilisation of the orbital environment since, as physics teaches us, two objects cannot occupy the same space. So, to avoid wrecking precious space assets, we need to assign slots and stick to them.

Proper orbital slot allocation is needed to achieve some key goals:

1. Spectrum Management: satellite communication systems require the allocation of radiofrequency spectrum to ensure interference-free operation. The radio spectrum can be effectively managed by assigning specific frequencies and orbital slots to different satellite systems, minimising interference between satellite networks.

2. Coverage and Capacity: by positioning satellites in specific slots, operators can ensure reliable coverage over specific geographic regions and maximise the data transmission capabilities of their systems. Imagine a flashlight shining down. The spot is brighter in the centre, and although some light also falls on the sides, it loses efficiency. Regarding satellites, the principle is the same, but using radio communications instead of light.

3. Inter-Satellite Coordination: proper slot allocation allows for the efficient coordination of spacing and separation distances between satellites, reducing the risk of signal interference and potential collisions. Two or more satellites can share a "traffic lane" as long as they keep a safe distance from each other.

4. Geostationary Orbit Utilisation: Effective allocation of geostationary orbital slots ensures equitable access to this limited resource and optimises its utilisation. The Universe might be infinite, but the space around Earth and its good spots are limited.

The allocation of orbital slots involves a coordinated effort among various stakeholders, including international regulatory bodies, national administrations, satellite operators, and frequency coordination organisations. The process typically includes the following steps:

1. International Coordination: International coordination is facilitated by organisations such as the International Telecommunication Union (ITU). The ITU manages the global allocation of radiofrequency spectrum and satellite orbital slots, ensuring the equitable distribution of resources among member states. The spectrum is allocated on a first-come, first-served basis. Once a demand is made, the spectrum must be used within a specific period.

2. Geographical Filing: Satellite operators submit filings to the ITU, indicating their desired orbital slots and frequency assignments. These filings include technical information about the satellite system, operational parameters, and the desired coverage area.

3. Frequency Coordination: In collaboration with regional frequency coordination organisations, the ITU reviews the filings and coordinates the frequency assignments to minimise interference between satellite systems.

4. Regulatory Approval: Once the filings are reviewed and frequency coordination is completed, national administrations grant regulatory approvals to satellite operators, allowing them to operate their satellites within the assigned orbital slots.

Orbital slot allocation is not without its challenges. The demand for orbital slots far exceeds the available resources, especially in prime locations such as geostationary orbit. This scarcity poses challenges in allocating slots to meet the growing number of satellite systems.

Coordinating the placement of satellites within orbital slots requires careful consideration to avoid interference and ensure sufficient spacing between systems. Achieving coordination among different satellite operators, especially in congested orbits, can be complex.

To add to the issues, as the number of satellites in orbit continues to increase, long-term sustainability becomes a concern. The efficient management of orbital slots is crucial to minimise space debris creation and preserve the usability of critical orbits for future generations.

But, like in the case of parking spots in the cities, the good orbital slots are few. This created a rush to occupy them, thus creating an uneven playing field between states with space-launching capability and those without it. Even for space-faring states, regions like LEO are becoming crowded. This is why, with all the slot allocation processes, we sometimes need to take measures to avoid collisions.

Collision Avoidance and Maneuver Planning

In the dynamic and congested realm of space, collision avoidance and manoeuvre planning are critical components of space traffic management.

Collision avoidance begins with precise orbit determination and continuously tracking objects in space. Space surveillance networks, such as the United States Space Surveillance Network (SSN) and other international tracking systems, monitor the positions and trajectories of satellites and space debris. Although humanity is capable of tracking very small objects, elements like the atmospheric conditions, speed and rotation of the object can make this tracking less precise. Therefore, to increase safety, the danger of collision is not calculated for the exact size of the object but considering a sort of virtual box around it. When two boxes intersect, the risk is higher.

Conjunction analysis involves assessing the risk of a collision between two space objects. It considers the objects' orbital parameters, predicted trajectories, and positional uncertainties. By conducting detailed risk assessments, satellite operators can determine the probability of a collision and take appropriate preventive measures.

If a potential collision is detected, satellite operators initiate manoeuvre planning to avoid the predicted collision. Manoeuvres involve altering the satellite's orbit through small thruster firings or more significant orbital adjustments. These manoeuvres aim to create sufficient separation between the satellite and the object of concern, minimising the collision probability. Performing collision avoidance manoeuvres consumes valuable onboard fuel. Satellite operators must carefully manage fuel usage to maximise the operational lifespan of satellites while maintaining collision avoidance capabilities throughout their mission.

Collision avoidance in space presents unique challenges due to the vastness of the orbital environment and the large number of objects in motion.

The time available to respond to a potential collision is often limited, especially in cases where objects are on intersecting trajectories. Satellite operators must swiftly calculate and execute appropriate manoeuvres before the predicted conjunction occurs.

As pointed out above, despite advanced tracking systems, there can still be uncertainties in the orbital data of satellites and debris objects. These uncertainties make accurate prediction of collision risks challenging, requiring conservative assumptions to ensure safety.

Effective collision avoidance relies on timely communication and coordination among satellite operators, space agencies, and space traffic management centres. Clear protocols and procedures for information exchange and coordination are necessary to facilitate efficient decision-making and prevent conflicting manoeuvres.

International collaboration plays a crucial role in collision avoidance efforts. Remember, a collision needs two parties, so preventing it generally also requires two parties to talk to each other. Imagine the situation where two cars are about to collide, and both drivers try to avoid it by steering towards the same point. It doesn't end well. Sometimes, there are more than two parties; sometimes, they do not have enough open communication lines, so we need to have common rules.

Space Surveillance and Tracking

Space surveillance and tracking technologies play a crucial role in space traffic management by providing essential information about the position, trajectory, and identification of space objects.

Space surveillance and tracking systems serve as the "eyes" of space traffic management, enabling the continuous monitoring of satellites, space debris, and other objects in orbit. They enable the identification and cataloguing of space objects, including operational satellites, defunct satellites, rocket stages, and fragments of space debris. Accurate identification is crucial for tracking and distinguishing between objects to assess collision risks and facilitate coordination among satellite operators.

By continuously tracking the positions and velocities of space objects, surveillance and tracking systems contribute to precise orbit determination. This data is essential for predicting potential close approaches between objects and calculating the risk of collisions. Accurate orbit determination allows timely collision avoidance manoeuvres and helps satellite operators make informed decisions.

The tracking capabilities of surveillance systems are particularly valuable in monitoring and characterising space debris as their data helps in understanding debris population dynamics, identifying debris mitigation strategies, and assessing the risks associated with space debris.

Several technologies are employed in space surveillance and tracking to monitor the orbital environment, including:

• Radar Systems that use radio waves to detect and track space objects. They emit pulses of radio frequency energy and measure the time it takes for the signal to bounce back from the object. By analysing the returned signals, radar systems can determine the object's distance, velocity, and position.

• Optical Systems, such as telescopes and cameras, capture images of space objects to track their positions and movements. These systems detect reflected sunlight or light emissions from the objects themselves. Optical tracking provides valuable data for precise orbit determination and object identification.

• Laser Ranging involves sending laser pulses towards retro-reflectors placed on satellites or space debris objects. By measuring the time it takes for the laser light to bounce back, the distance to the object can be accurately determined. Laser ranging offers high-precision tracking capabilities, especially for specific objects with retro-reflectors.

• Radio Frequency (R.F.) Monitoring involves tracking the radio signals emitted by satellites. The satellite's position and velocity can be determined by analysing the frequency, intensity, and Doppler shift of these signals. R.F. monitoring is particularly useful for tracking satellites with active communication systems.

Effective space surveillance and tracking require collaboration and integration of multiple tracking systems worldwide. International organisations share data and collaborate with countries and organisations to achieve a comprehensive and globally coordinated surveillance network. Integrating data from various systems improves orbit determination accuracy, facilitates timely collision warnings, and enhances the overall space traffic management capabilities.

However, the deep mistrust between countries has led to independent systems that do not work together, a lack of communication lines and deep suspicion of advancing cooperation and regulatory frameworks.

Debris Mitigation Measures

Debris mitigation measures are essential for minimising the creation of new space debris and reducing the risk of collisions in Earth's orbit. These measures encompass a range of practices and guidelines that satellite operators and launch providers follow to limit debris generation during satellite missions and promote the long-term sustainability of space activities.

One of the key debris mitigation measures is the implementation of post-mission disposal procedures. Satellite operators are encouraged to remove their satellites from operational orbits at the end of their mission to minimise the risk of collisions and allow for the reuse of that orbit. This involves moving the satellite to a graveyard orbit, where it is less likely to interfere with operational satellites or deorbiting it to burn up safely in the Earth's atmosphere. Proper disposal of satellites reduces the accumulation of debris in valuable orbital regions.

Satellites and space systems can be designed with the "design for demise" principle in mind. This means incorporating materials and components that are more likely to burn up during re-entry into the Earth's atmosphere, thereby reducing the amount of debris that reaches the ground. Designing satellites for controlled re-entry ensures that they disintegrate and disperse harmlessly, minimising any potential risks associated with their re-entry.

To prevent the creation of additional debris through collisions or intentional actions, operators take measures to minimise the fragmentation of satellites or rocket stages. This includes ensuring structural robustness to withstand impacts, utilising less prone to fragmentation materials, and implementing collision avoidance measures to reduce the risk of collisions with other space objects.

Furthermore, innovative propulsion systems and autonomous operations enable satellites to manoeuvre to safe orbits or deorbit at the end of their missions, further reducing the risks associated with space debris.

While debris mitigation measures focus on preventing the creation of new debris, active debris removal (ADR) is an emerging technology to remove existing space debris from orbit. ADR missions involve capturing and deorbiting defunct satellites, spent rocket stages, or other large debris objects that pose significant collision risks. This technology holds great promise in reducing the density of debris in critical orbital regions and mitigating the long-term debris hazard. Still, the regulatory efforts have hit a wall as countries are wary of the potential development and use of military strategies and technologies.

Space debris mitigation measures are supported by international guidelines and best practices. These guidelines, such as those provided by the Inter-Agency Space Debris Coordination Committee (IADC) and the International Organization for Standardization (ISO), outline specific recommendations for satellite operators and launch providers. Compliance with these guidelines ensures a standardised approach to debris mitigation and promotes responsible space practices globally.

Raising awareness about the importance of debris mitigation is crucial in promoting responsible behaviour among satellite operators, launch providers, and space agencies. Educating stakeholders about the risks and consequences of space debris and the best practices for debris mitigation can foster a culture of responsible space operations. This includes providing training and resources to satellite operators, encouraging technology development for debris tracking and removal, and promoting international collaboration to address the debris challenge collectively.

By implementing effective debris mitigation measures, we can proactively manage the space debris environment, reduce collision risks, and safeguard the long-term sustainability of space activities, thus ensuring a cleaner and safer space environment for future generations.

Information Sharing

Today, more than twenty countries can launch things into space, a significant increase from the three countries that could do so during the intense space race in the 1960's. Additionally, many other national and local groups have their own satellites. As technology improves and the costs of space launches decrease, even more countries are expected to join the ranks of those exploring space.

Space Traffic Management is mainly a challenge related to governing space activities rather than a technical problem. To operate safely in space, those in charge need to work together. This involves coordinating and communicating with each other, sharing data and information, staying aware of what's happening in space, and finding ways to avoid conflicts. It also means creating processes and procedures to decide who will make certain moves in space, when they'll make those moves, and how they'll do it to keep everyone safe.

Currently, the management of space objects is a bit informal and not well-coordinated. For strategic and even military reasons, some information is not communicated, and states keep their cards close, risking incurring damages.

However, as space becomes more crowded and complicated, many experts and leaders call for better international governance of space traffic.

Sharing information is vital to space traffic management because it allows different groups, like satellite operators and space agencies, to work together, coordinate their activities, and make informed decisions. This is crucial to ensure that space remains a safe and sustainable place for exploration.

CHAPTER SEVEN

The Path to Harmony

Over the years, humanity has come together to create space treaties aimed at establishing a responsible and peaceful framework for the exploration and use of outer space.

These treaties are designed to prevent the militarization of space, reduce the risk of conflicts, and promote cooperation in the peaceful use of celestial bodies and outer space. They reflect a shared commitment to fostering a cooperative and inclusive approach to space exploration, ensuring that the benefits of space activities are shared for the betterment of humanity. For example, the Outer Space Treaty of 1967 outlines principles that prohibit weapons of mass destruction in space and national appropriation of celestial bodies.

The United Nations and the Outer Space Treaty

The primary framework governing space activities is the Outer Space Treaty of 1967, a foundational agreement outlining the principles of space exploration. However, the rapid growth of space traffic has prompted the development of additional guidelines. The United Nations Committee on the Peaceful Uses of Outer Space (COPUOS) plays a central role in fostering collaboration among spacefaring nations to address challenges such as debris mitigation, space situational awareness, and the registration of space objects.

In the context of space debris governance, the Outer Space Treaty outlines the principle of environmental protection in outer space. According to this treaty, states must adhere to the principle of environmental protection in their outer space activities and prevent harmful contamination of outer space, including the Moon and other celestial bodies. This did not stop US astronauts from leaving 96 bags of human waste on the Moon, but let's assume nothing harmful grows there.

Additionally, the treaty mandates states to authorise and continuously supervise the space activities of non-governmental entities. This oversight also extends to commercial space activities, contributing positively to preventing space debris generation.

The Registration Convention complements these principles by requiring the registration of space objects launched into outer space. This registration provides a legal basis for determining the responsible entity for space debris governance. However, while these treaties establish fundamental principles, specific rules regarding space debris governance are predominantly found in "soft law" documents or mechanisms.

The Space Debris Mitigation Guidelines, adopted by the Inter-agency Space Debris Coordination Committee (IADC) in 2002, represent a significant step. The IADC, an intergovernmental international organisation, coordinates global space activities related to debris issues. Their guidelines establish a comprehensive debris mitigation mechanism, covering prevention during normal operations, on-orbit break-up minimisation, post-mission disposal, and on-orbit collision prevention. Each space project must prepare a feasible space debris mitigation plan, addressing debris mitigation management, risk assessment, malfunctions hazard minimisation, and end-of-mission disposal.

Another document titled Space Debris Mitigation Guidelines, adopted by the United

Nations Committee on the Peaceful Uses of Outer Space (UNCOPUOS) in 2007, builds upon the IADC guidelines. They summarise the sources of space debris and establish prevention mechanisms for spacecraft at different stages. Countries with spacefaring capabilities are urged to establish domestic mechanisms for space debris mitigation. The UNCOPUOS plays a crucial role as the main platform for strengthening space debris governance and fostering international recognition of relevant rules.

The UNCOPUOS Guidelines on the Long-term Sustainability of Outer Space Activities (LTS Guidelines), established in 2019, contain 21 specific guidelines. These include strengthening the registration of space objects and sharing space debris monitoring information. However, contentious issues, like active removal of space objects and collision-avoidance measures, remain unresolved and subject to ongoing discussions.

The transparency and confidence-building measures in outer space activities (TCBM), initiated by the U.N. General Assembly in 2011, aim to formulate consensus recommendations on outer space transparency. Improving transparency in space activities is crucial for effective space debris governance, especially in space information exchange and situational awareness. The TCBM mechanism requires sharing information about space objects and related events, establishing data processing and sharing systems, and fostering international cooperation.

Due to escalating security and environmental risks associated with space debris, recent years have seen new trends in developing international rules for space debris governance. Firstly, there are growing calls for establishing legally binding rules governing space debris management. Delegations at the UNCOPUOS STSC conference in 2021 proposed creating such binding rules to institutionalise and systematise space debris governance further.

As commercial entities become prominent players in space exploration, the need for transparent and standardised regulations becomes even more pressing. The prospect of mega-constellations, satellite servicing missions, and space tourism necessitates collaborative efforts to avoid congestion and potential conflicts in Earth's orbit.

New Policies and Regulations

As the debris problem is international, and in the framework of the United Nations the progress is slow, many spacefaring countries have developed their own policies and regulations, but not all of them include the issue of space traffic management or space debris. Some of the more relevant ones can be found below, but keep in mind that there are currently 43 countries that have national legislation on space activities, and 77 countries are participating in U.N. works on space, but very few tackle the issue of space debris.

Since 1988, the U.S. has emphasised minimising orbital debris creation, termed orbital debris mitigation. The 2010 National Space Policy outlines the significance of protecting the space environment and mitigating orbital debris: "Orbital debris poses a risk to continued reliable use of space-based services and operations and to the safety of persons and property in space and on Earth. The United States shall seek to minimise the creation of orbital debris by government and non-government operations in space in order to preserve the space environment for future generations."

In June 2018, the National Space Traffic Management Policy, Space Policy Directive-3 (SPD-3) stated: "Orbital debris presents a growing threat to space operations. Debris mitigation guidelines, standards, and policies should be revised periodically, enforced domestically, and adopted internationally to mitigate the operational effects of orbital debris. ... The United States should develop a new protocol of standard practices to set broader expectations of safe space operations in the 21st Century. This protocol should begin with updated ODMSP but also incorporate sections to address operating practices for large constellations, rendezvous and proximity operations, small satellites, and other classes of space operations. These overarching practices will provide an avenue to promote efficient and effective space safety practices with U.S. industry and internationally."

Often blamed for the uncontrolled re-entry of space debris and an anti-satellite weapon test, China took policy steps in this matter. In 2015, the China National Space Administration (CNSA) established the Space Debris Monitoring and Application Center, an entity responsible for monitoring, early warning, emergency response, and international cooperation related to space debris. Various regulations and policy documents have been issued, including the Measures for the Administration of Registration of Space Objects (2001) and the Interim Measures on Space Debris Mitigation and Protective Management (2009). The Outline of the Space Debris Action

Plan from 2006 to 2020 and national technical standards like Space Debris Mitigation Requirements (GB/T 34513).

The White Paper on China's Space Activities 2016 highlighted the country's aims for the space debris problem, as did the following white paper, China's Space Program: A 2021 Perspective. This includes strengthening space traffic control, enhancing the space debris monitoring system and cataloguing database, and improving early warning services. Additionally, the plan incorporates in-orbit maintenance of spacecraft, collision avoidance and control, and space debris mitigation, all aimed at ensuring space systems' safe, stable, and orderly operation. The existing documents related to space debris management mainly consist of departmental regulations or policy documents. However, there are plans for a comprehensive National Space Law.

In the world almost 30 countries have national laws aimed at space activities, out of which 11 are in Europe.

Recently, the European Space Agency launched the non-legally binding Zero Debris Charter, Towards a Safe and Sustainable Space Environment.

The European Commission, the body in charge of proposing new European Union legislation, has also announced that a EU Space Law (EUSL), which will cover space traffic management and sustainability, is scheduled to be presented in 2024.

The EUSL is expected to comprise three pillars that are aimed at achieving the following objectives:

1. Ensure the safety of satellite traffic by addressing the increasing risk of collisions and damages caused by space debris.

2. Coherently protect the EU and national space infrastructures and assets against harmful threats, particularly against cyberattacks.

3. Guarantee the long-term sustainability of space operations, ensuring the EU's ability to depend on space as a significant enabler of services and economic growth.

Clearing the haze

Legal texts are creating haze zones in outer space, where the rules of governance and conduct are unclear or disputed.

There is no clear definition of "peaceful" use or purposes in international space law, and military activities in outer space remain largely ungoverned. Some countries are unwilling to accept limitations on their use of space. There are many uncertainties, particularly in rapidly changing activities and technologies.

Countries have always used all elements of their national power, including military, economic, and alliances, to deter or stop aggression. Currently, humanity is neither at war nor at peace, and space is not an absolute vacuum.

Article 48 of the ITU constitution allows states complete freedom to use military radio on allocated frequencies.

The UN Registration Convention is not effectively applied to military activities. Very few satellites are registered as having military functions; information on their actual uses and capabilities is scarce, and their number is not exact.

Satellites require ground infrastructure, computer systems, and end-user terminals to operate, and connections are not solely physical. The constant stream of data from satellites to Earth gives us television, radio, telecommunications, and cyberspace, but also helps military force deployment, national security and defence decision-making. Earth activities also affect space, and satellites can be hacked, causing severe effects across the space system.

The biggest challenge to casting light on the haze is that this haze is not solely an effect of governance but also a component of governance. Great power's competition and interests make new top-down governance efforts challenging, particularly in international bodies such as the United Nations.

Emerging Technologies and Innovations

The legal future of space traffic management is closely intertwined with developing and implementing emerging technologies and innovative solutions.

In space traffic management, automated collision avoidance systems can prevent potential collisions between objects by rapidly detecting possible close approaches. These systems can identify collision risks and provide timely warnings by analysing real-time orbital data and employing sophisticated algorithms. The ability to respond quickly is crucial in avoiding collisions, as even minor trajectory adjustments can significantly help prevent close approaches.

These systems can accurately calculate the probability of collisions by considering factors such as uncertainties in orbital data, atmospheric drag, and gravitational perturbations. This allows for precise manoeuvre planning to avoid potential collisions.

Automated collision avoidance systems operate 24/7, continuously monitoring the space environment. They track the positions and velocities of space objects, update orbital data, and assess potential collision risks in real-time.

This constant monitoring ensures a proactive approach to collision avoidance and allows for prompt response to any emerging threats.

By automating the collision avoidance process, these systems reduce the burden on satellite operators, who would otherwise have to manually analyse data and plan avoidance manoeuvres. Automated systems can perform complex calculations and simulations, optimising trajectory adjustments and minimising fuel consumption. This improves operational efficiency and extends the lifespan of satellites.

In response to the increasing complexity of space traffic management, the concept of space traffic coordination constellations is gaining traction. These constellations consist of dedicated satellites or sensors strategically positioned in orbit to monitor and track space objects. They are not exactly the equivalent of traffic lights in space or some space beacons, but they help manage traffic. By providing comprehensive and continuous coverage of the space environment, these constellations enhance space situational awareness, improve collision prediction accuracy, and facilitate timely coordination of space operations. Space traffic coordination constellations can serve as a global monitoring network, enabling effective management of satellite constellations, debris tracking, and collision avoidance efforts.

Developing advanced sensors and tracking technologies also plays a vital role in space traffic management. High-resolution optical sensors, radar systems, and other advanced tracking technologies enable more precise measurements and improve orbit determination, better identification and tracking of space debris, and enhanced space situational awareness. Integrating multiple sensor types, such as optical, radar, and radio frequency, further improves the accuracy and reliability of tracking data, enabling more effective collision prediction and avoidance.

In-orbit servicing (IOS) and Active Debris Removal (ADR) technologies offer innovative solutions to address the space debris menace and ensure the long-term sustainability of space activities. IOS capabilities involve repairing, refuelling, and repositioning satellites in orbit. By extending the operational lifespan of satellites, IOS reduces the need for new satellite deployments and minimises space debris generation. ADR technologies, on the other hand, aim to actively remove defunct satellites and debris from orbit. These technologies include capture and deorbiting mechanisms, such as robotic arms, nets, or harpoons, that can safely remove debris from critical orbits, reducing the collision risks for operational satellites.

Given the rapid development of large satellite constellations, establishing unified technical standards is essential. These standards should cover launch, deployment, rendezvous and proximity operations, collision avoidance, and disposal of constellation satellites. There is a lack of unified technical standards for space rendezvous and proximity operations, which are crucial for effective space debris management. Establishing integrated and unified technical standards becomes particularly relevant for addressing debris issues arising from constellation operations. Assessing the orbital space occupied by specific satellite constellations is also essential to ensure the rational and equitable use of orbit and frequency resources.

Improving constellation information sharing and early warning mechanisms is equally vital. Large constellations, including component satellites, are anticipated to occupy a significant portion of orbit space. Sharing constellation information promotes space information exchange, while early warnings enable timely traffic management and collision avoidance, reducing the risks associated with information asymmetry and lack of communication in outer space activities.

CHAPTER EIGHT

Our Cosmic Legacy

As we look ahead to the future, the destiny of humankind is interwoven with the vast expanse of space. Our journey as a spacefaring species opens new chapters in exploration, discovery, and collaboration.

Imagine a future where humans live and work in space permanently. Permanent space stations will serve as our homes in orbit around Earth. These stations are not just for astronauts conducting experiments; they're like mini-cities where scientists, engineers, and everyday people can live and contribute to the ongoing space exploration. They'll be platforms for learning, innovation, and understanding our place in the cosmos.

Our closest neighbour in space, the Moon, is beckoning us to establish bases. These lunar bases will act as stepping stones, helping us learn more about the challenges of living on other celestial bodies. Just as pioneers explored new lands on Earth, but more responsible, we will venture beyond our planet, turning the Moon into a hub for research, resource extraction, and launching missions to even more distant destinations.

Mars, the mysterious and captivating Red Planet, promises to become a second home for humanity. In the future, we might build colonies there, where people can live, work, and build a society. Which laws will they apply?

A thrilling new frontier is asteroid mining. Imagine sending robotic spacecraft to asteroids, extracting valuable minerals, and returning them to Earth or other space stations. This will fuel our exploration and provide access to rare and precious resources that can sustain our growing population and support future space missions. But who will own those resources? And what if two countries or companies target the same spot?

As we extend our reach into the cosmos, it's crucial to uphold the principles of peace and cooperation. International agreements must guide our actions and prevent the militarisation of space. Only by fostering diplomacy and open communication can we ensure that our exploration of the cosmos is a collaborative and harmonious endeavour.

Sustainability is the key to a lasting cosmic legacy. Responsible practices in space, such as minimising space debris and developing sustainable and ecological technologies, will preserve the integrity and accessibility of the space and celestial bodies. By actively participating in debris removal initiatives and adopting green space technologies, we become stewards of the cosmos, ensuring that the wonders of space endure for generations to come.

Standing at the threshold of this cosmic future, the possibilities are as limitless as the stars in the night sky. Through the collective efforts of scientists, engineers, explorers, and dreamers, we are writing the next chapters of our cosmic story.

BIBLIOGRAPHY

V. Mandl, Das Weltraum-Recht. Ein Problem der Raumfahrt, J. Bensheimer Verlag, Mannheim, Berlin, Leipzig, Germany, 1932

E. Pépin, Les problèmes juridiques de l'espace, The McGill Law Journal. 6, 1959

Treaty on Principles Governing the Activities of States in the Exploration and Use of Outer Space, including the Moon and Other Celestial Bodies (1967)

Agreement on the Rescue of Astronauts, the Return of Astronauts and the Return of Objects Launched into Outer Space (1968)

Convention on International Liability for Damage Caused by Space Objects (1972)

Convention on Registration of Objects Launched into Outer Space (1976)

Agreement Governing the Activities of States on the Moon and Other Celestial Bodies (1984)

IADC Space Debris Mitigation Guidelines, IADC-02-01, Revision 1, Inter-Agency Space Debris Coordination Committee, 2007

United States National Space Policy, 2010

White Paper on China's Space Activities 2016

E.S. Nightingale, B. Lal, B. Weeden, A.J. Picard, A.R. Eisenstadt, Evaluating Options for Civil Space Situational Awareness (SSA), Institute for Defence Analyses, Science and Technology Policy Institute, Washington, DC, 2016.

Space Policy Directive-3 (SPD-3) National Space Traffic Management Policy, United States (2018)

S. Moranta, T. Hrozensky, M. Dvoracek, Towards a European Approach to Space Traffic Management, European Space Policy Institute, Vienna, Austria, 2020.

China's Space Program: A 2021 Perspective

D. Oltrogge, M. Strah, M. Skinner, R.J. Rovetto, A. Lacroix, A.A. Kumar, K. Grattan, L. Francillout, I. Alonso, Recommendations of the IAF Space Traffic Management Terminology Working Group, in: Dubai, UAE, 2021.

Space Traffic Management -A brief history, Quentin Verspieren LSE IDEAS, 2022

The Government Accountability Office, Large Constellations of Satellites GAO-22-105166 (2022)

Jessica West and Jordan Miller, Clearing the Fog: The Grey Zones of Space Governance, CIGI Papers No. 287, 2023

Space Industry Debris Mitigation Recommendations, World Economic Forum, 2023.

European Commission, ec.europa.eu (accessed 2023)

NASA Orbital Debris Program Office (accessed 2023)

Zero Debris Charter - Towards a Safe and Sustainable Space Environment, ESA 2023

European Space Agency, esa.int (accessed 2023)